D0929585

Introduction to Industrial Chemistry

Introduction to Industrial Chemistry

HOWARD L. WHITE
CIBA-GEIGY Corporation (Ret.)

A Wiley-Interscience Publication

JOHN WILEY & SONS
New York • Chichester • Brisbane • Toronto • Singapore

Library of Congress Cataloging in Publication Data:

White, Howard L., 1922-
Introduction to industrial chemistry.

"A Wiley-Interscience publication."
Includes bibliographies and index.
1. Chemistry, Technical. I. Title.

TP145.W46 1986 660.2 86-5643
ISBN 0-471-82657-X

10 9 8 7 6 5 4 3 2 1

To my wife Betty,
whose encouragement was vital in my
writing this book
and whose help in the detailed assembly
of the manuscript
was essential.
To my daughter, Joyce
(who suggested a word processor)
and to my sons Howie and Chris.

Acknowledgment

It is with pleasure that I acknowledge the insight into industrial chemistry gained during my career at Ciba-Geigy Corporation from my association with many colleagues on a wide variety of projects.

The use of the facilities of the Brown University Sciences Library has been invaluable to me.

HLW

Preface

A deficiency has developed in recent years in undergraduate chemistry programs taken by students heading toward a career in some branch of the chemical industry. To remedy the situation, students should enroll in a one-semester course in industrial chemistry using this textbook. This should be taken during their third or fourth year of study, after they have completed a general, an analytical, and an organic chemistry course. This book clarifies the career areas and technical problems to be considered when chemical reactions are carried out on a large scale. My twenty-two years in industrial chemical process development and experience in teaching industrial chemistry at the university level have been used in compiling this material.

Since World War II, understanding of chemistry fundamentals has increased at an extremely rapid rate. Undergraduate chemistry programs have consequently become much more theoretical. At the same time, a large number of consumer products based on chemical processes have come on the market. The research and development necessary for discovering these laboratory-scale processes and their conversion to an industrial scale have been carried out by industrial laboratories, but have usually not been described in university chemistry courses. New employees in the chemical industry have not been adequately prepared to deal with the parameters within which the chemical industry operates. This book should help with the indoctrination stage of their jobs.

The rapid increase in the volume of theoretical chemical data that undergraduates must cover has forced a reduction in the amount of descriptive chemistry taught, because of time limitations. However, many scientists now feel that too much of this kind of material has been removed from the curriculum. I wrote this book to give the basic background needed to handle a selected group of large-scale organic, fermentation, and inorganic processes. This type of material is necessary for a student who wishes to be adequately trained as a chemist. This book discusses these processes with modern mechanistic concepts in mind. This approach will help to hold the interest of the student who does not like highly theoretical treatments.

There has developed a distinct antiscience feeling among young people in recent years. For example, the preparation of chemical industry products such as pesticides and herbicides is seen as having caused severe pollution problems. This textbook discusses these types of problems by giving examples of how a product can be prepared in high yield with correspondingly low effluent. Bioconversion of product and effluent to innocuous products is discussed. In summary, the book discusses attack by chemical and biological means on these problems.

The all-important economic aspects of a process are emphasized in this book. Simple cost calculations to pinpoint factors that contribute to cost and to show how much it costs for pollution and effluent cleanup are explained. Selection of methods for cleanup at minimum cost is clarified.

This text does cover many monomers but does not get into polymers. The author believes polymer chemistry should be a separate one-semester course taken following this one. The principles of scale-up, cost, pollution, and so on discussed here will be readily applied in polymer work.

I believe that training in current B.S., M.S., and Ph.D. chemistry programs is strongly research oriented. Graduates are not generally aware that industrial chemistry has a wide range of careers that have not been mentioned in chemistry major programs. Also, discussions of chemical processes are usually omitted. I recommend enriching advanced courses through use of industrial material. This would have the result that students and faculty would get a more solidly based view of what to expect in various industrial careers in research and development, and in other careers needing a chemical background.

I also believe that industrial chemistry in the classroom should not be narrowly focused on vocational considerations, but should be taught to a much larger segment of the student population than just the industry-bound chemist.

Industrial chemistry should be a distinct portion of the curriculum, not just supplementary notes and comments. The difficulty is that academics lack appreciation of industrial concerns, and also that industrial chemists scold their academic colleagues for not teaching practical chemistry.

Industrial chemistry should be included in the curriculum of all students who are considering a chemical career. The material is of equal importance to that in the current course lists.

HOWARD L. WHITE

Warwick, Rhode Island
June 1986

Contents

Introduction to Industrial Chemistry

1

The Chemical Industry and Large-Scale Chemical Manufacturing

We commence by describing those substances prepared by the group of companies that make up the chemical industry. We group these products according to how the public uses them in order to direct the reader's attention to those materials used in everyday living that originate from industrially manufactured chemicals. The many chemical process steps needed to show the connection between some of these useful substances and their raw materials are described. There are two general aims in view.

First, it is necessary to show how the reactions studied in previously taken chemistry courses have to be adapted in preparing these commonplace substances on a large scale. A discussion is presented to illustrate what problems arise when the scale of the process increases beyond what can be done in the laboratory up to the preparation of commercial quantities. Examples of these factors are regulation of environmental damage and safety and economic considerations.

Second, we discuss these processes so that university graduates will have an introductory background for careers such as state and federal regulators, or other careers affected by the preparation of these products in enormous quantities, as well as for chemical industry jobs.

The chemical industry's product groups used in the monograph *Chemistry in the Economy* (Harris and Tishler, 1973, pp. vii–viii) are as follows:

Fertilizers
Ferrous metals
Glass
Other inorganic chemicals
Petroleum refining products
Natural and synthetic rubber
Plastics and resins
Textile fibers

Soaps and detergents
Protective coatings
Pharmaceuticals
Personal care products
Pesticides
Food processing chemicals
Pulp and paper chemicals
Photographic products
Electronic equipment products
Electrical equipment products
Other organic chemicals
Products of the nuclear industry

A few selected products from the preceding categories have been chosen for detailed discussion. Some are inorganic products and some are organic chemicals. Others could have been used just as easily to illustrate the principles to be discussed. The aim is to lead you to think about the details of *any* chemical process by seeing how these particular materials are prepared.

The previous laboratory experiences of most chemistry students have probably only included the preparation of, at the most, a hundred grams of a selected group of chemicals. The lecturer may have referred to a few equations for some industrial processes. This book shows the many opportunities for the use of a chemist's background knowledge in the development and scale-up stages for preparing a large-scale process for a useful chemical product. Merely having prepared in the laboratory a substance that has been shown to have an exciting end use does not mean that the large-scale stage will be easily reached.

1.1 BACKGROUND FACTORS IN LARGE-SCALE PROCESSES

In this chapter we introduce various topics relating to large-scale processes that are not considered at all in laboratory work at most universities. The syntheses, at least the equations, may well have been encountered in other courses. In some cases the process chemistry was discovered many years ago, but necessary improvements are still being made because of questions of cost, environmental damage, and so on.

1.1.1 Raw Materials

It probably does not occur to the typical student that all the chemicals in the university laboratory have a basic natural source, for example, a mineral, petroleum, an agricultural crop, or an atmospheric gas. When millions of kilograms of a product are to be prepared, it is important to determine the availability of the raw material in the area slated for manufacture. All prospective chemists should realize that only a small number of chemicals made for use as laboratory reagents are prepared in commercial quantities. When a source of raw material for a new process is being considered, it is necessary to think about the limited supply of our earth's resources, which, of course, controls the quantities of basic source material available. Then, to limit the quantity of raw material needed, organic and inorganic process development chemists have to become involved in recycling products from waste streams, conversion of chemical by-products from a process to make additional useful materials, and the devising of one process to make two commercially useful products at once. Background of this sort is discussed in Chapters 4, 5 and 6.

Both inorganic and organic raw materials are used in sequential reactions in which substances are converted to products that are sold, or, in other cases, converted to different commercial products. Chenier (1983) shows this in tabular form for some inorganic chemicals existing in nature. His data are shown in adapted form in Figure 1.1.

These data indicate that certain substances on the list of the commodity chemicals (those produced in enormous quantities) are produced from naturally

Nature's Raw Materials	Other Raw Materials		Products	
Air		→	$O_2 + N_2$	(1)
$CH_4 + H_2O$		→	$CO + H_2$	(2)
N_2 (as air) +	H_2	→	NH_3	(3)
O_2 (as air) +	NH_3	→	HNO_3	(4)
O_2 (as air) + S + H_2O		→	H_2SO_4	(5)
Al_2O_3 +	H_2SO_4	→	$Al_2(SO_4)_3$	(6)
$CaF_2 \cdot 3[Ca_3(PO_4)_2]$ +	H_2SO_4	→	H_3PO_4	(7)
$NaCl + H_2O$		→	$NaOH + Cl_2 + H_2$	(8)
	$H_2 + Cl_2$	→	$2HCl$	(9)
Alkanes +	Cl_2	→	Cl-alkanes + HCl	(10)
$CaCO_3$		→	$CaO + CO_2$	(11)
$CaCO_3$ + NaCl		→	$CaCl_2 + Na_2CO_3$	(12)
SiO_2 +	Na_2CO_3	→	$Na_2O \cdot SiO_2$	(13)
	$CO_2 + NH_3$	→	NH_2CONH_2	(14)
TiO_2 (as air) +	$Cl_2 \rightarrow TiCl_4$	→	TiO_2	(15)

FIGURE 1.1 Summary of preparations of several commodity chemicals. From P. J. Chenier, *J. Chem. Ed.* **60**, 412 (1983). Reprinted with permission.

occurring materials. The "other" raw materials are those previously prepared from natural raw materials in a separate synthesis as shown in this figure.

The numbers in the following list refer to Figure 1.1.

1. Oxygen and nitrogen are prepared by the distillation of liquid air.
2. Syngas (a mixture of carbon monoxide and hydrogen) is made from methane (or other hydrocarbons) with steam. (See Section 2.3.1.A, B.)
3. Ammonia is synthesized from nitrogen and hydrogen. (See Section 2.3.2.)
4. Nitric acid is prepared from ammonia by catalytic oxidation with oxygen. (See Section 2.5.)
5. Sulfuric acid is made from sulfur, oxygen and water in a three-step process. Sulfur is mined or recovered from petroleum. (See Section 2.4.)
6. Aluminum sulfate is prepared from sulfuric acid and aluminum oxide (alumina), which is mined.
7. Phosphoric acid is prepared from phosphate rock, a mineral called fluoroapatite, a calcium fluoride–calcium phosphate double salt, and sulfuric acid. (See Section 2.1.)
8. Soduium hydroxide and chlorine are prepared by the electrolysis of sodium chloride solution. (See Section 2.6.)
9. Hydrogen and chlorine combine directly to give hydrogen chloride.
10. Hydrogen chloride is more commonly made from the reaction of various alkanes (from petroleum) with chlorine. The chloroalkanes are discussed in Section 4.8.3.C.
11. Lime is made from the mineral limestone by heating.
12. Sodium carbonate (soda ash) is made from the minerals limestone and salt with the aid of ammonia and coke. This is the solvay process. It is also mined as a component of the mineral Trona.
13. Sodium silicate (glass) is made from sand, lime, and soda ash.
14. Carbon dioxide reacts with 2 moles of ammonia to give ammonium carbamate, which decomposes to urea and water.
15. Titanium dioxide, a product used in paints, is made from the mineral rutile (TiO_2), which is purified by distillation of the tetrachloride.

To obtain a similar outlook on organic raw materials, see Chapters 3 and 4.

1.2 REVIEW OF MAJOR TOPICS

We now examine in synopsis form the highlights of the book:

Type of process discussed. Examples of inorganic, fermentation, and organic processes are examined. Chemically simple processes have been included, along with a few more elaborate ones.

Scale-up. The techniques used in changing a laboratory process to large scale are discussed in Chapter 5. These are problems that are of little or no importance to the laboratory chemist.

Costs. As we consider a process, we need a procedure for deciding which are technically the most difficult factors so as to properly direct our research and development work. The student would probably think about problems strictly related to the laboratory, such as long reaction time and wasteful recrystallization. On a large scale, other problems could be, for example, the use of an expensive piece of equipment, or a difficult-to-prepare raw material, or troublesome waste disposal. To decide where to aim the development work, a simplified cost calculation is prepared. The method used is given in Chapter 6. The most technically difficult stage is shown as the most costly step in this preliminary process computation. Future process work will then be directed toward simplifying the process, with resulting cost reduction.

Energy Requirements. In the laboratory, electric heaters are used for supplying energy and ice baths are used for removing heat. For large-scale work, it is necessary to keep energy costs as low as possible. For example, it is sometimes technically straightforward to transfer energy emitted by a reaction to another process step where it is needed. Energy costs are reduced by the use of catalysts and by utilizing heat of reaction in various ways. (See Chapters 2, 8 and 11.)

Chemical Control. It is necessary to have chemical analyses done at various points in a process and to understand the chemistry of the product and of the by-products that is occurring at each stage. This is necessary to determine quickly when the yield is at the maximum, and to minimize the quantity of effluent and other factors. (See Chapters 2, 4, and 5.)

Use of Catalysts. The great majority of large-scale industrial processes, particularly the continuous ones, require catalysts. These substances are used to cause other substances to react quickly with reduction or elimination of solvent and by-product and with resulting cost reduction. From the needs of the chemical industry, particularly the petrochemical industry, new synthetic reactions have been achieved using special catalyst mixtures. These reactions are rather unusual synthetic steps. The impetus for discovering these reactions was the preparation of products having great industrial importance at minimum cost. This book briefly discusses some detailed mechanistic work used to plan the most cost effective final process. (See Chapters 2, 4 and 11.)

Effluent. Each process where very large quantities are prepared has to be carried out at over 90% yield when cost and environmental factors are considered. Even then the effluent presents a considerable disposal problem: It has to be analyzed, usable products have to be recovered, and the rest must be disposed of in an environmentally safe way. This is discussed in Chapter 7.

Equipment. Chapter 8 covers the equipment to be discussed when the chemist changes over from a lab reactor, usually a round bottom flask, to large-scale vessels. Among the considerations are whether the process is a continuous or batch process, the material of construction, preparing equipment for effluent treatment, and studying necessary time cycle changes.

1.3 ASSIGNMENTS TO BE COMPLETED DURING THE COURSE

To understand the material presented in the course, two projects are strongly recommended as assignments.

1. The student should prepare a detailed term paper on one chemical process, using published technical data.
2. The class should visit several chemical manufacturing plants in the area. One plant visit should be chosen for the assembly of a trip report. Instructions for preparing these reports follow.

1.3.1 How to Prepare the Term Paper

The term paper should consist of a detailed description of any chemical process carried out on an industrial scale. The aim of this exercise is to ensure that the student uses what has been learned in a practical way. For instance, a student might cover a process in a company with which he or she is considering employment. The student would choose any procedure in which a considerable amount of data is available from technical or patent sources. The principles discussed in class lectures should be applied to the preparation of the paper. When data for a particular process stage are not published, students are encouraged to do library research and suggest answers.

The paper should cover the following:

1. Description of the product's use and the quantity manufactured yearly.
2. A brief description of the process—one paragraph.
3. Chemical equations for all process steps.
4. The raw material sources, including a short statement on the basic raw material source and how it is converted to the chosen raw material.

5. Analyses needed on raw material, to determine whether the quantity of by-products is within acceptable limits. Only a list of procedures is needed.
6a. An equipment flow sheet, showing items needed: for example, 1000-L vessel, filter, drier, numbered so that the flow of materials can be followed.
6b. A step-wise description of the process.
7. A list of analyses needed to tell when the reaction is complete and to follow the reaction.
8. A list of analyses of the product for purity and presence of significant by-products.
9. Specifications for the product.
10. Effluent from process—gas, liquid, solid. What processing is needed before dumping. Suggest ways to decrease quantity of effluent.

1.3.2 How to Prepare the Trip Report

An important part of the course consists of visits to chemical manufacturing plants. A trip report is needed covering *one* of the plants visited. It should consist of a short account of the student's own observations of the plant. The report should show that he or she understands the way the particular plant functions and also the chemistry of the process. It should not be just a copy of the instructor's material or company handout.

Items to be included are the following:

1. Description of the quality and source of the raw material used.
2. Quantity of product produced from a given quantity of raw material (plant may have some restrictions on this information).
3. Description of the equipment and the chemical process at the particular plant. Show how closely process control is achieved, how desired quality of product is produced, and what sort of analytical control on effluent is carried out.
4. The student's own impression should be given on the trapping of by-products and their disposal.

Throughout the report, tables, diagrams, and chemical equations should be used, rather than wordy descriptions.

1.3.3 Nomenclature

The writer has stated that certain chemistry courses should have been taken prior to using this book. It is hoped that these have included up-to-date instruction on nomenclature. Industrial people have regrettably not kept up with this trend to modern names. The author has used the chemical names he thinks are in current use in industry in 1986 in the United States. New employees should be alert to this problem, and be prepared for the use of two names for the same chemical compound in the trade literature.

REFERENCES

Chenier, P. J., *J. Chem. Educ.*, **60**, 412 (1983).

Harris, M., and M. Tishler, *Chemistry in the Economy*, American Chemical Society, Washington, D.C., 1973, pp. vii–viii.

QUESTIONS

These questions are to be answered using only the information in this chapter and your own background knowledge. They are meant to focus your attention on what the book covers.

1. What branch of the chemical industry is of direct concern in your daily life? (one paragraph)
2. What chemical raw materials are produced close to your area?
3. How many chemical industry products do you think are really essential to *your* life today? What community problems do you think are caused by their manufacture? What can be done about the problems?
4. What is your impression of how the chemical industry looks after its waste? How should this problem be tackled?
5. Discuss any experiment carried out in a previously taken organic chemistry course, and describe what problems you think will be encountered in the manufacture of 1 million kilograms of the product.

2

Inorganic Processes for Study

Major plant nutrients necessary for worldwide agricultural fertilizer needs are nitrogen, phosphorus, and potassium. This book discusses eight processes, six involved with fertilizer production, one for high-purity phosphoric acid, and one for sodium hydroxide and chlorine:

Phosphoric acid
Superphosphate
Triple superphosphate
Ammonia
Sulfuric acid
Nitric acid
High-purity H_3PO_4 (for food additive use)
Sodium hydroxide and chlorine

2.1 WET PROCESS FOR PHOSPHORIC ACID AND SUPERPHOSPHATES

Bone chips and organic wastes have been used as a phosphorus agricultural supplement for many years. In 1840–1842, von Liebig in Germany and John B. Lawes in England found that treating bones with sulfuric acid made the nutrients more available to plants. The raw material for all phosphorus products is phosphate rock, some of which probably originated from the bones of animals who lived millions of years ago. The formula most used for this is $CaF_2 \cdot 3\,[Ca_3(PO_4)_2]$, which is the raw material called fluoroapatite. This raw material is produced in various African countries, the U.S.S.R., and the United States in very large quantities.

For phosphorus-containing fertilizers, we now use superphosphate and triple superphosphate, which are calcium acid salts of phosphoric acid, as described below, made by partial neutralization of the phosphate rock. Phosphoric acid is prepared mainly for making various salts for fertilizer use and also in highly purified form for manufacture of salts as food additives. A review of this material is available (Childs, 1977). These processes are reviewed below.

2.1.1 Processes for H_3PO_4 Manufacture

Two processes are available, both in commercial use.

A. *The Wet Process (for Fertilizer-Grade H_3PO_4)*

First we discuss the neutralization of phosphate rock with sulfuric acid.

$$CaF_2 \cdot 3[Ca_3(PO_4)_2] + 10H_2SO_4 + 20H_2O \rightarrow$$

$$\underset{\text{Gypsum}}{10CaSO_4 \cdot 2H_2O\downarrow} + \underset{\text{Phosphoric Acid}}{6H_3PO_4} + 2HF\uparrow$$

The following additional reactions occur because

1. There is SiO_2 in the raw rock.
2. There is water present in the scrubber, reacting with the SiF_4 as shown:

$$4HF + SiO_2 \rightarrow SiF_4 + 2H_2O$$

$$3SiF_4 + 2H_2O \rightarrow 2H_2SiF_6 + SiO_2$$

The $CaSO_4 \cdot 2H_2O$ is separated by filtration. The resulting acid is impure and is used in the manufacture of fertilizer.

B. *The Thermal Process (for High-Purity H_3PO_4)*

The thermal process is a three-stage process. The phosphorus, prepared by the reduction of phosphate rock, is purified by distillation. This purified material is burned to P_4O_{10} which is added to water to yield phosphoric acid.

$$CaF_2 \cdot 3[Ca_3(PO_4)_2] + 9SiO_2 + 15C \rightarrow$$

$$9CaSiO_3 + CaF_2 + 6P + 15CO \quad \text{(I)}$$

$$2CaF_2 + 3SiO_2 \rightarrow SiF_4 + 2CaSiO_3$$

$$3SiF_4 + 2H_2O \rightarrow 2H_2SiF_6 + SiO_2$$

$$4P + 5O_2 \rightarrow P_4O_{10} \quad \text{(II)}$$

$$P_4O_{10} + 6H_2O \rightarrow 4H_3PO_4 \quad \text{(III)}$$

The resulting acid is a highly purified product suitable for conversion to salts for use in foodstuffs. The process is discussed in Section 2.2.

C. *Discussion of Wet Process (for Phosphoric Acid)*

The process is written to inform the chemist who has had only laboratory experience and the procedure is presented as an industrial chemist sees it. The acid and prepulverized rock are intimately mixed, and sufficient time for digestion (complete reaction) is allowed. Special equipment is required to prepare large quantities of very finely ground phosphate rock. The $CaSO_4 \cdot 2H_2O$ (gypsum) has to be precipitated in easily filterable form. Obtaining the correct conditions for the crystallization is a key point in the process.

Procedure. This is a continuous process. For background on this type of process see Chapter 8 (particularly Figure 8.5.*e*,*f*), and for further details on this particular process see Kulp and Leyshon (1968). Figure 2.1 shows how the Dorr–Oliver version of this process is carried out.

1. For the neutralization stage, there is a single-tank reactor (1), fitted with baffles and agitators as shown. To (1) is continuously fed a stream of finely

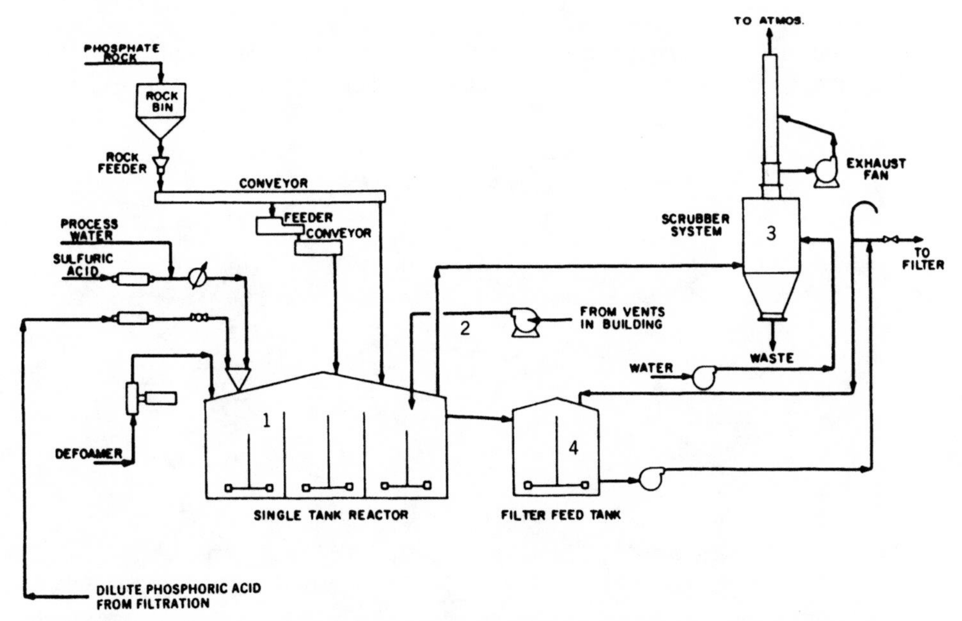

FIGURE 2.1 Equipment diagram for phosphoric acid manufacture—wet process. (Reprinted with permission from R. L. Kulp and D. Leyshon, *Phosphoric Acid*, A. V. Slack, Ed., Vol. 1, Part I. Marcel Dekker, New York, 1968, p. 215.)

ground phosphate rock, and some recycled phosphoric acid mother liquor along with sulfuric acid. A current of air pumped in at (2) causes cooling by evaporation. While the phosphoric acid and calcium sulfate dihydrate are forming in the digestion process, the gas SiF_4 is evolving and being trapped in the scrubber (3), reacting with the water as shown in Section 2.1.1.A.

2. The slurry is sent to the filter feed tank (4), and then is continuously filtered on the Bird–Prayon continuous filter, a set of tilting-pan filter boxes rotating as shown in Figure 2.2. As explained in Shreve and Brink (1977, p. 254), the feed continuously enters the pans, which are connected to the vacuum source. The circular frame supporting the pans rotates so that each pan is moved successively under the desired number of washes. After the final wash liquor has completely drained off, the vacuum is released and the pan is inverted a full 180° to dump the cake. The filter is cleaned, and the pan is returned to the feed position. Washing and dumping of the cake portions is continuous. The strong acid mother liquor is recycled to (1) and the weak wash liquid is used to wash the earlier cakes. The $CaSO_4 \cdot 2H_2O$ is filtered on the slowly moving pans. The phosphoric acid is the undiluted mother liquor.

Product: an aqueous H_3PO_4, [analysis 30–32% P_2O_5 (41–44% H_3PO_4)].

Cost Aspects of Process. The following are some of the factors increasing cost:

1. If a more concentrated acid is needed, water is evaporated from the acid. Evaporation is an expensive step in any product of low cost. Attempts are

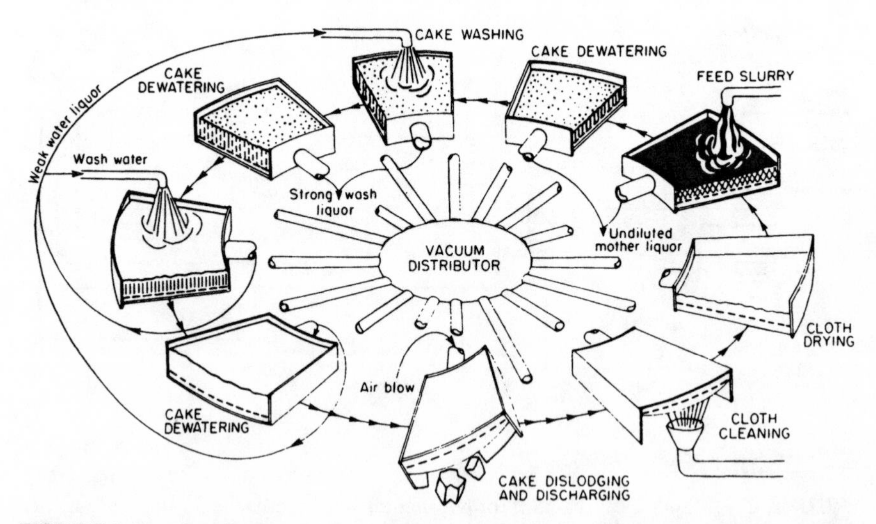

FIGURE 2.2 Phosphoric acid manufacture—wet process. Continuous filtration equipment for removing gypsum. (From Bird Machine Co., Inc.; Neponset St.; S. Walpole, MA 02071.)

being made to use a more concentrated sulfuric acid for mixing with the phosphate rock to obtain a more concentrated phosphoric acid. (Childs, 1977, pp. 388–389). A 64% sulfuric acid is now used.

2. Maintenance of correct crystallization conditions for $CaSO_4 \cdot 2H_2O$. If $CaSO_4$ with one-half mole of water of crystallization is formed (called plaster of paris, the form used in making casts for broken bones), the lines containing the slurry to be filtered may plug.
3. Purchase and installation of a continuous filter system is costly.
4. Noncorrosive equipment to stand silicon tetrafluoride and hydrogen fluoride attack is needed.

Energy. Not much input of energy is needed, unless concentration of the acid is required.

Chemical Controls. The following are the chief control factors:

1. It is necessary to check whether as much as possible of the rock has been converted.
2. Maintenance of correct conditions for $CaSO_4 \cdot 2H_2O$ formation is needed. Control of temperature and of acid concentration and seeding with recycled slurry as well as a well-baffled tank reactor are among the factors to be watched.
3. Measurement of SiF_4 evolution and control of corrosion is necessary.
4. Many phosphoric acid plants use another method of cooling the slurry in (1). This is to pump a continuous fraction of the slurry to a cooler, then back to the tank reactor. The air-cooling system described above seems to be as efficient and the process proceeds just as well (Kulp and Leyshon, 1968).

Effluent. Safe disposal must be provided for the following materials:

1. Fluoride neutralization products from scrubbing systems. In recent years some of the fluosilicic acid has been used as a raw material for various fluorine chemicals.
2. Gypsum ($CaSO_4 \cdot 2H_2O$) is formed in large quantities. It has a commercial use but too much gypsum results from this process to be worth preparing for sale.

2.1.2 Superphosphate

The original work of Lawes and von Liebig on treating bones with sulfuric acid gave a monocalcium phosphate $CaH_4(PO_4)_2$. This acid salt is more soluble in water than is powdered bone, and therefore more available to the plant. This product is called superphosphate.

$$CaF_2 \cdot 3[Ca_3(PO_4)_2] + 3H_2O + 7H_2SO_4 \rightarrow$$

$$3CaH_4(PO_4)_2 \cdot H_2O + 2HF + 7CaSO_4 \cdot 2H_2O$$

This process is similar to that for wet process phosphoric acid manufacture. The rock is only partially neutralized. The gypsum is not removed. The ground rock is reacted with the acid on a slowly moving belt in a "den," as shown in Figure 2.3. As can be seen from the diagram, the acid rock mass slowly moves through the den while reaction takes place and then is chopped and placed on a curing pile to permit the reaction to be completed. The evolved HF is trapped in a scrubber, reacting with water as shown in Section 2.1.1.A. The reaction is complete after 10–20 days. The mixture contains all the material added except for the hydrogen fluoride that was evolved. It is called superphosphate and consists of a calcium diphosphate–calcium sulfate mixture.

Product: Superphosphate, a granular solid [analysis 16–20% P_2O_5 (22–28% H_3PO_4)].

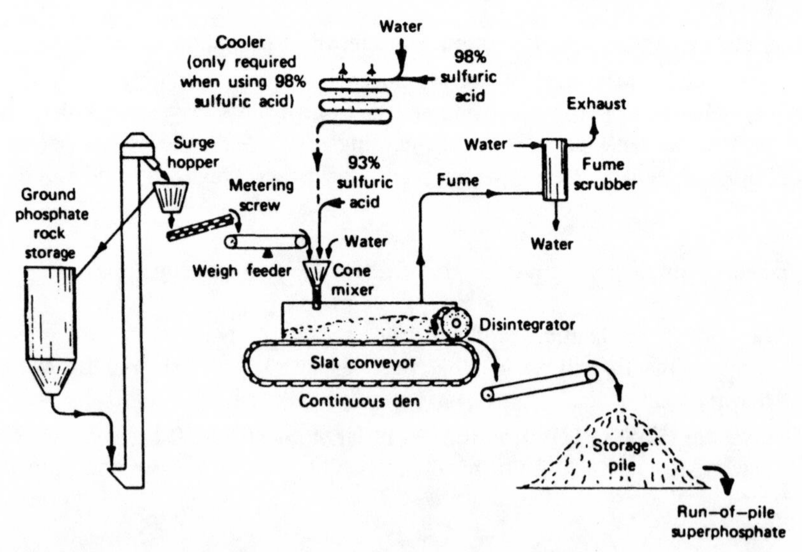

FIGURE 2.3 Flow chart for superphosphate—continuous den process. (Reproduced with permission from E. O. Huffman, *Encyclopedia of Chemical Technology*, 3rd ed., R. Kirk and D. Othmer, Eds. Wiley, New York, 1980, Vol. 10, p. 73).

2.1.3 Triple Superphosphate

The process and equipment are the same as for superphosphate, except that neutralization is carried out with phosphoric acid:

$$CaF_2 \cdot 3[Ca_3(PO_4)_2] + 14H_3PO_4 \rightarrow 10CaH_4(PO_4)_2 + 2HF$$

The product is ground. Clearly, no filtration is needed.

Product: A granular solid [analysis 44–51% P_2O_5 (61–70% H_3PO_4)]. The product is more concentrated than normal superphosphate because it contains no $CaSO_4$.

2.2 THERMAL PROCESS FOR PHOSPHORIC ACID

The equations for this process are given in Section 2.1.1.B. They show a reduction of the rock to elemental phosphorus, which is purified by distillation. The very high reduction temperature requires an electric furnace. The phosphorus is burnt to give P_4O_{10}, which is hydrated to H_3PO_4.

2.2.1 Procedure Details

The procedure is as follows (Childs, 1977, pp. 378–382):

Stage I (see Figure 2.4*a*)

1. Phosphate rock, coke, and silica are smelted in a closed electric furnace (A) at 1550°C. A mixture of carbon monoxide and phosphorus vapor is driven off the top of the furnace, as well as gaseous fluoride by-products (see below). The CO is recycled and burned as an energy source for preheating phosphate rock entering the autoclave.
2. Slag (calcium silicate and ferrophosphorus) is taken off below as shown (B).
3. Phosphorus vapor is condensed in water sprays (C). It is stored under water. The water has the products from the HF (see Section 2.1.1.A).

Stage II (see Figure 2.4*b*)

1. A spray of molten phosphorus is burned in air and steam in a water cooled burner made of stainless steel.
2. Exit gases are passed into a hydrator, where phosphoric acid is formed, which is sent to the strong acid tank. The remaining gases are sent to a venturi and a cyclone to remove mist.

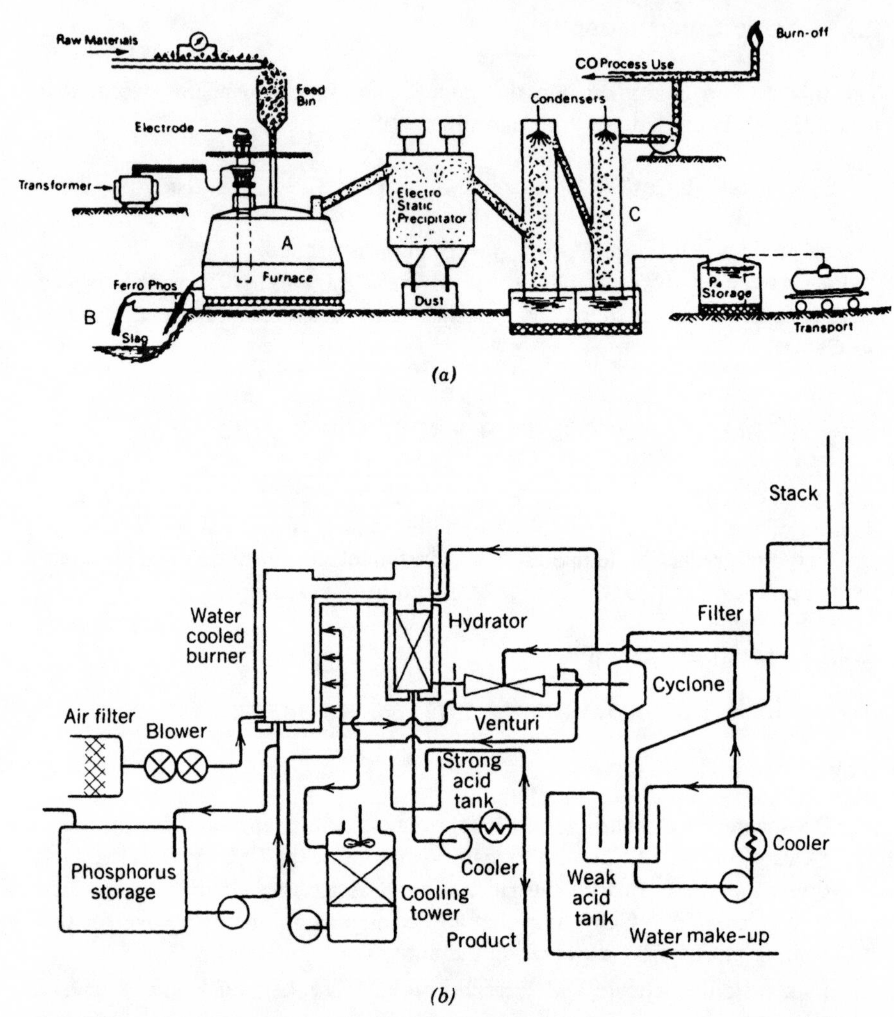

FIGURE 2.4 (a) Elemental phosphorus manufacture for preparing thermal process phosphoric acid. Flow chart. (b) Phosphoric acid manufacture—thermal process. Flow chart. (Reproduced with permission from A. F. Childs, in *The Modern Inorganic Chemicals Industry*, R. Thompson, Ed. Copyright by Royal Society of Chemistry, 1977, pp. 380, 386.)

Product: An aqueous H_3PO_4, [analysis 62% P_2O_5, (85–86% H_3PO_4)] (material more concentrated than that from the wet process).

2.2.2 Process Notes

It is necessary to keep in mind the following factors in reviewing this process:

A. *Chemical Controls*

The following aspects require precise chemical control:

1. Removal of phosphorus in water effluent from the phosphorus process.
2. Control of arsenic in phosphorus (Childs, 1977, p. 382) (As <136 ppm).

B. *Cost Aspects*

The reduction step uses a great deal of energy. It is economical only where energy is inexpensive, for example, in Newfoundland, Canada, where there is a phosphorus manufacturing unit using this procedure.

C. *Effluent*

The following comments show how various effluents are handled (Childs, 1977, p. 382):

1. The aqueous runoff from the process is highly toxic to fish if it contains even traces of phosphorus compounds. The water is neutralized, lime or ammonia is added, and the mixture is allowed to settle. Then, to remove the last traces of phosphorus, a chlorine treatment is given.
2. The slag (mostly calcium silicate) can be used as concrete aggregate.
3. Ferrophosphorus is used in the manufacture of special steels and cast irons.

2.3 AMMONIA

Nitrogen is usually supplied for agricultural use in the form of ammonia or ammonium salts. The same result can be achieved by the use of Chilean nitrates and by crops that have a nitrogen-fixing ability.

Ammonia is an inefficient system for supplying nitrogen to plants. The absorbed NH_3 is oxidized to nitrate, with the loss of hydrogen which was costly to introduce into the ammonia molecule. A commercially viable process for fixing nitrogen as nitrate does not exist.

2.3.1 The Haber Process for Ammonia

A review of the Haber process is supplied by Andrew (1977, pp. 201–221).

Ammonia is made from nitrogen gas from air and hydrogen from synthesis gas (often called syngas).

$$N_2 + 3H_2 \rightarrow 2NH_3$$

A. Synthesis Gas Process Summary

The term *synthesis gas* is used for gas mixtures consisting of carbon monoxide and hydrogen in various proportions. This mixture was originally generated from coke with air and steam. Oil or natural gas can also be used instead of coke, depending on what $H_2:CO$ ratio is required. The $H_2:CO$ ratio for the resulting synthesis gas is 1:1 for coal, 2:1 for oil, 2.4:1 for gasoline, and 4:1 for methane (natural gas). To simplify the chemistry of synthesis gas formation, we discuss two processes.

Steam Hydrocarbon Reforming. The process has two stages.

Stage 1 is an endothermic process. Equations are given for various alternative raw materials: carbon, methane, or butane. These are mixed with steam at high temperature.

$$C + H_2O \rightarrow CO + H_2$$

$$CH_4 + H_2O \rightarrow CO + 3H_2$$

$$C_4H_{10} + 4H_2O \rightarrow 4CO + 9H_2$$

The above equations are examples of the general reaction:

$$C_nH_m + nH_2O \rightarrow nCO + (0.5m + n)H_2$$

Stage 2 is an exothermic process. This equation is for the so-called shift reaction

$$CO + H_2O \rightarrow CO_2 + H_2$$

Partial Oxidation. This is an exothermic process represented by the following reactions:

$$CH_4 + 2O_2 \rightarrow CO_2 + 2H_2O$$

$$CH_4 + CO_2 \rightarrow 2CO + 2H_2$$

$$CH_4 + H_2O \rightarrow CO + 3H_2$$

The overall reaction from these three processes is

$$CH_4 + 0.5O_2 \rightarrow CO + 2H_2$$

B. Industrial Processes for Preparation of Synthesis Gas

To clarify the sort of variation in procedure that has been used in industry for synthesis gas formation, three processes are briefly described.

Using coal, we have the following processes:

1. *Winkler* uses fine-grain nonbaking coals, O_2 or air plus steam at atmospheric pressure, and temperature in the 800–1100°C range. The product has a H_2:CO ratio of 1.4:1.

2. *Koppers–Totzek* uses powdered coal gasified at 1400–1600°C. High temperature eliminates partially converted hydrocarbons (85–90% conversion to products).

3. *Lurgi* uses lumpy coal or briquets, degassed at 20–30 bar and 600–700°C and fed downward against O_2 fed upward.

C. Details of Haber Process for Ammonia

As you read the process description that follows. Figure 2.5 should be consulted.

In *vessel 1* the natural gas is hydrogenated over a Cr–Mo catalyst. The evolved hydrogen sulfide is absorbed in zinc oxide. The desulfurized gas is passed to *vessel 2A*, the first of two steam reforming vessels (2A and 2B). These contain nickel oxide catalyst. This is activated by the hydrogen from the thermal cracking of higher hydrocarbons in natural gas.

$$NiO + H_2 \xrightarrow{\text{Cr–Mo}} Ni + H_2O$$

In vessel 2A the following endothermic reaction occurs:

$$CH_4 + H_2O \xrightarrow{\text{Ni}} CO + 3H_2$$

This is a process requiring heat input through reactor walls to displace the reaction to the right. This high temperature brings about a high pressure, which shifts the equilibrium to the left. In practice, the temperature is kept relatively low (700–800°C) and the pressure at about 400 psig, and the resulting gas mixture of H_2, CO, CO_2, and unconverted CH_4 is sent to *vessel 2B*, the secondary steam reformer. Air is admitted to the vessel, and methane is converted to carbon monoxide and hydrogen by the partial oxidation exothermic process.

$$CH_4 + \text{air} \xrightarrow[\text{940°C}]{\text{Ni}} CO + 2H_2 + N_2$$

Because of this it operates at a higher temperature than vessel 2A, even with no heat input through vessel walls. The gas mixture of N_2, H_2, CO_2, CO, and CH_4 (0.2%) is one example of a syngas with a particular CO:H_2 ratio. (Syngas

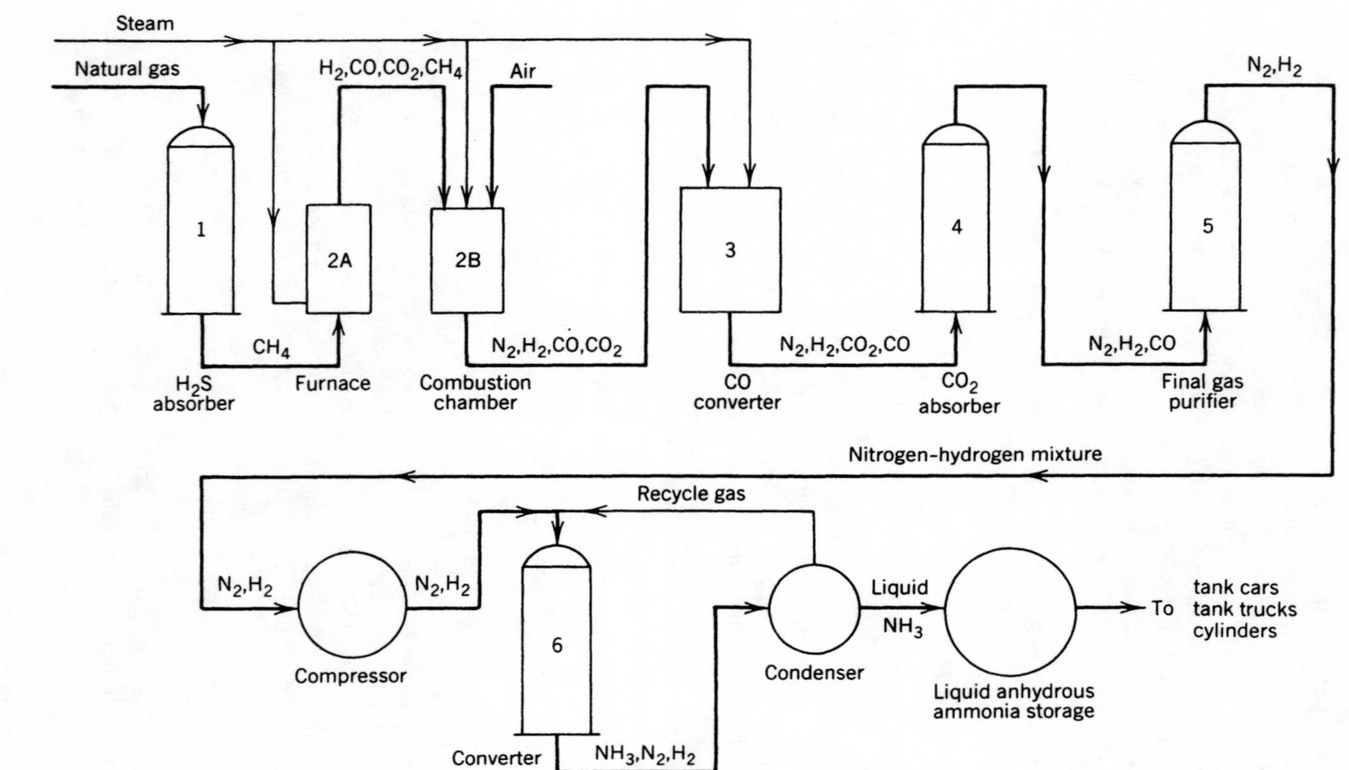

FIGURE 2.5 Flow chart for ammonia manufacture by steam reforming of natural gas. (Material obtained through the courtesy of C-I-L, Inc., North York, Ontario.)

is made with various ratios depending on the several syntheses using this mixture) (see Chapter 4). It is sent to *vessel 3*. The water gas shift reaction occurs, reducing the carbon monoxide from 10% to 0.2%.

$$CO + H_2O \xrightarrow[400°C]{FeO-Cr_2O_3} CO_2 + H_2$$

Then the gas mixture flows to *vessel 4*. Here, monoethanolamine (MEA) is used to remove CO_2 from the process. The CO_2 can be recovered by reboiling the MEA·CO_2 solution, yielding CO_2 for synthesis (see below) and MEA for absorbing more CO_2. To use the recovered CO_2, a new process has been developed for a synthesis gas process where the H_2:CO ratio is satisfactory for methanol production. Ammonia and methanol plants are integrated as described in Wittcoff and Reuben (1980, p. 116).

$$3CH_4 + CO_2 + 2H_2O \xrightarrow{Ni} 4CO + 8H_2$$

$$2H_2 + CO \xrightarrow{Cu} CH_3OH$$

See Section 4.2.5.A. After CO_2 is removed, the gases are sent to *vessel 5*. Removal of the last traces of CO (less than 10 ppm CO) is necessary to prevent poisoning of the ammonia synthesis catalyst. This step is called *methanation* and takes place at 315°C over a nickel catalyst.

$$CO + 3H_2 \xrightarrow{Ni} CH_4 + H_2O$$

$$CO_2 + 4H_2 \longrightarrow CH_4 + 2H_2O$$

The gases are then sent to *vessel 6*, in which the Haber process for ammonia is carried out:

$$N_2 + 3H_2 \rightarrow 2NH_3$$

The catalyst of choice is metallic iron promoted by KOH and containing small amounts of Al_2O_3, SiO_2, and MgO. Haber showed that high pressure would be necessary for a satisfactory rate of reaction and yield. Even then the conversion is low. The ammonia is removed by refrigeration in a synthesis loop that recycles unconverted nitrogen and hydrogen. The need for large compressors in vessel 6 limits the size of the converter. Large compressors are very expensive. The pressures needed are 80–350 atm.

A new idea for avoiding the use of large compressors is to generate hydrogen at a high pressure (da Rosa, 1978). This is achieved by carrying out the electrolysis of water in a suitably compact electrolysis unit mounted inside a pressure vessel.

The reaction is temperature as well as pressure sensitive (Andrew, 1977, p. 209). The best temperature range is 400–540°C. Below 400°C, the catalyst does not bring about an adequate rate, and above 540° its surface area is lost because it sinters.

2.4 SULFURIC ACID

The production of sulfuric acid in the United States in 1984 was 79.37 billion pounds. It is the largest volume chemical produced. About 70% of this product is used in fertilizer manufacture. For most uses, the concentration should be 91% or greater. To achieve this required concentration, the contact process was devised.

2.4.1 The Contact Process for Sulfuric Acid

A review of the contact process is presented by Phillips (1977 pp. 185–200), and Bland (1984). The raw material for this procedure is either elemental sulfur, sulfur dioxide, or pyrite (FeS_2). The sources of these materials are as follows:

1. Sulfur from mines (by Frasch process)
2. Sulfur or hydrogen sulfide recovered from petroleum desulfurization
3. Recovery of sulfur dioxide from coal or oil-burning public utility stack gases
4. Recovery of sulfur dioxide from the smelting of metal sulfide ores:

$$2PbS + 3O_2 \rightarrow 2PbO + 2SO_2$$

5. Isolation of SO_2 from pyrite (only in eastern Europe)

2.4.2 Steps in the Contact Process

As the steps in this process are examined, Figure 2.6 should be consulted.

1. Burning of sulfur ($S + O_2 \rightarrow SO_2$) (vessel 1)
2. Catalytic oxidation of SO_2 to SO_3 (vessel 2)

$$2SO_2 + O_2 \rightarrow 2SO_3$$

3. Hydration of SO_3

$$SO_3 + H_2O \rightarrow H_2SO_4 \text{ (vessels 3A and 3B).}$$

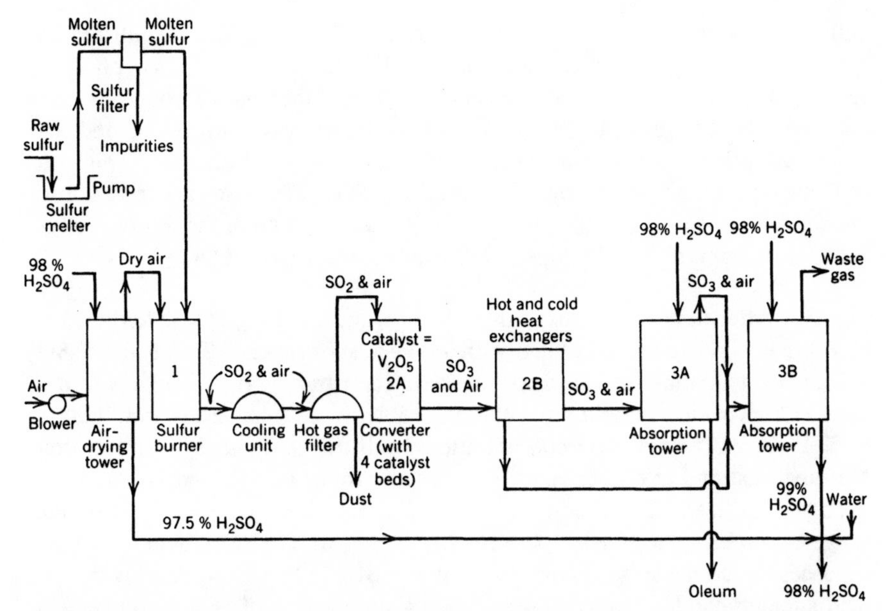

FIGURE 2.6 Flow chart for sulfuric acid manufacture by the contact process. Piping between catalyst layers from 2A to 2B, and also to 3rd unit (3) not shown, see Section 2.4.2A-Step 2. (Material obtained through the courtesy of C-I-L, Inc., North York, Ontario.)

A. Process Details (see Figure 2.6)

Step 1. Dry SO_2 is needed for the catalytic oxidation. With sources 1 and 2 above, dry air is used in the combustion process. Water will cause corrosion because of acid formation. If sulfur contains carbonaceous impurities, the molten material has to be filtered to avoid poisoning the catalyst and forming water from burning hydrogen.

Step 2. When using sulfur from sources 1 and 2, purification of the SO_2 gas is normally not needed. Other sources of SO_2 require wet scrubbing followed by treatment of the gas with electrostatic precipitators to remove fine particles. The catalyst used is vanadium pentoxide and the pressure is 1.2–1.5 atmospheres. The temperature has to be kept around 410–430°C. If it rises above 430°C, the equilibrium is displaced away from SO_3. The value has to reach around 410°C for the catalyst to be activated. This process is strongly exothermic. Consequently, the catalytic reactor (2A) is designed as a four-stage fixed-bed unit. The gas has to be cooled between each step. Four passes, together with "double absorption," described below, are necessary for overall conversion of 99.5–99.8% (three passes, 97–98%). The temperature rises to over 600°

with the passage of the gas through each catalyst bed. The doubled absorption consists of cooling the gases between each bed back to the desired range by sending them through the heat exchanger (2B), and then back through the succeeding beds. Between the third and fourth beds, the gases are cooled and sent to an absorption tower like unit (3). This is not shown on diagram. This is to shift the equilibrium to the right by absorbing SO_3. The gases are then sent to the heat exchanger to warm them to 410-430° and then on to the fourth catalyst bed. For a diagram, see Phillips (1977, p. 185 and Bland 1984).

Step 3. After the catalytic oxidation process, the resulting SO_3 is hydrated by absorption in packed towers filled with 98–99% sulfuric acid. This is the H_2SO_4 azeotrope of minimum total vapor pressure. The catalytic oxidation has to proceed in high yield to avoid air pollution problems. SO_2 has a low solubility in 98% H_2SO_4. At lower acid concentrations, sulfuric acid and SO_3 form a troublesome mist and at higher concentrations emissions of SO_3 and H_2SO_4 vapor become significant. The absorption acid concentration is kept within the desired range by exchange as needed between the H_2SO_4 in the drying acid vessel that precedes the combustion chamber with the H_2SO_4 in the absorption tower. The acid strength can be adjusted by controlling the streams of H_2SO_4 to give acid of 91 to 100% H_2SO_4 with various amounts of added SO_3 and water. The conversion of sulfur to acid is over 99.5%.

In a sulfuric acid plant much usable heat results from each of the three steps described earlier. Consequently, modern plants are coproducers of H_2SO_4 and steam or power. Also some heat of reaction is used to melt the sulfur.

B. *Mechanism of Vanadium Pentoxide Catalysis*

In a review by Emmett (1975), it is pointed out that the V_2O_5 catalyzed oxidation of sulfur dioxide is usually described as a heterogeneous catalysis reaction. Emmett points out that this is not correct. The current catalyst is promoted V_2O_5, and has added alkali metal sulfate. The catalyst is made up of an inert support (e.g., SiO_2), an alkali metal sulfate, and a source of vanadium. The first step is to activate the catalyst. Some SO_2 is passed over the material and a pyrosulfate salt is formed ($SO_2 \rightarrow SO_3$ by heterogeneous catalysis).

$$SO_4^{2-} + SO_3 \rightarrow S_2O_7^{2-}$$

At the normal reaction temperature for this oxidation, the pyrosulfate salt is a liquid, and is capable of dissolving vanadium salts. The catalyst as it reacts is a granule of silica covered with molten pyrosulfate in which some vanadium with oxidation state of 4 or 5 is dissolved. A mechanism for the process is

$$2V^{5+} + O^{2-} + SO_2 \rightarrow SO_3 + 2V^{4+}$$

$$2V^{4+} + 0.5O_2 \rightarrow 2V^{5+} + O^{2-}$$

If the temperature falls, the vanadium salt precipitates out, and the catalytic activity falls off. Also when the temperature rises to 640°C, it was seen that a gas mixture of SO_2 and O_2 produced a yield of only 60% SO_3. The rate-determining step is the solution of oxygen in the melt.

2.5 NITRIC ACID

Nitric acid is made by the oxidation of ammonia, using platinum or platinum–10% rhodium as catalyst, followed by the reaction of the resulting nitrogen oxides with water. This process has been reviewed by Andrew (1977, pp. 221–228). The hydrogen is introduced to the ammonia by reactions that need energy. The soil bacteria oxidize the ammonia to nitrate and water before it can be absorbed by the plant. It is clear that the energy used for the preparation of ammonia is wasted. We need a process for fixing nitrogen as nitrate. This does not as yet exist.

The reactions that are involved in the ammonia oxidation are as follows:

$$NH_3 + 1.25O_2 \rightarrow NO + 1.5H_2O \quad (1)$$

$$2NO + O_2 \rightarrow 2NO_2 \quad (2)$$

$$3NO_2 + H_2O \rightarrow 2HNO_3 + NO \quad (3)$$

Overall

$$NH_3 + 2O_2 \rightarrow HNO_3 + H_2O$$

Each of the preceding reactions evolves heat, but not a sufficient amount to be significant in ammonia production.

The process has to be run under conditions that have been worked out to minimize side reactions, which are as follows:

$$4NH_3 + 3O_2 \rightarrow 2N_2 + 6H_2O \quad (4)$$

$$4NH_3 + 6NO \rightarrow 5N_2 + 6H_2O \quad (5)$$

$$2NO \rightarrow N_2 + O_2 \quad (6)$$

Reaction (4) is noncatalytic and occurs on hot reactor surfaces before the gases reach the catalyst. The reaction can be controlled by keeping the walls at as low a temperature as possible. As far as reaction (5) is concerned, the yield has to be as high as possible so that there will be very little NO in the reagent ammonia and also little or no NH_3 in the gases after the catalyst. This result and the minimization of reaction (6) are achieved by having the catalyst in the form of fine gauzes, which keep holdup of the reactants to a minimum.

2.5.1 Brief Process Description

As you read this description consult Figure 2.7.

1. Into the ammonia–air mixer is fed vaporized filtered liquid ammonia and preheated filtered compressed air. This mixture is led to the converter.

2. The converter contains the catalyst gauzes. This is where reaction (1) occurs. As shown in Figure 2.7, the gases leaving the converter are NO with enough oxygen for reaction (2), as well as nitrogen from the air and from reaction (6), and water mainly from reaction (1) (temperature, 920°, pressure, 1–4 atm). This is to compress gases because of equipment size limitations. Pressure does not affect the reaction as there is little volume change; see Eq. (1). The gases are cooled.

3. The oxidation of NO to NO_2 takes place after the gases leave the cooler, as well as the absorption in water to give nitric acid. Note the line from the bottom of the condenser for dilute nitric acid. This is from the water from reaction (1) with the NO_2.

4. The absorption tower has enough water so that 60% nitric acid will result, and also enough air to convert all remaining NO to NO_2. The pressure is 10 atm, to give more efficient absorption.

A more concentrated acid is obtained by distillation. The yield is 94–95% nitric acid. The product is 60% HNO_3. An acid of up to 65% strength can be obtained if the absorbing system is cooled to 2°C.

2.6 SODIUM HYDROXIDE AND CHLORINE

The two substances sodium hydroxide and chlorine are produced in one reaction by the electrolysis of brine (saturated sodium chloride). There are two sources of salt. It is produced by evaporation of sea water and it is mined in areas such as southeastern Michigan near Detroit.

There are three electrolysis processes. The diagrams in Figure 2.8 are basic to the diaphragm and the membrane cells. The processes have been reviewed by Shreve and Brink (1977, pp. 214–221) and Purcell (1977, pp. 117–120). In Figure 2.8*a*, the anode and cathode compartments are separated by a diaphragm

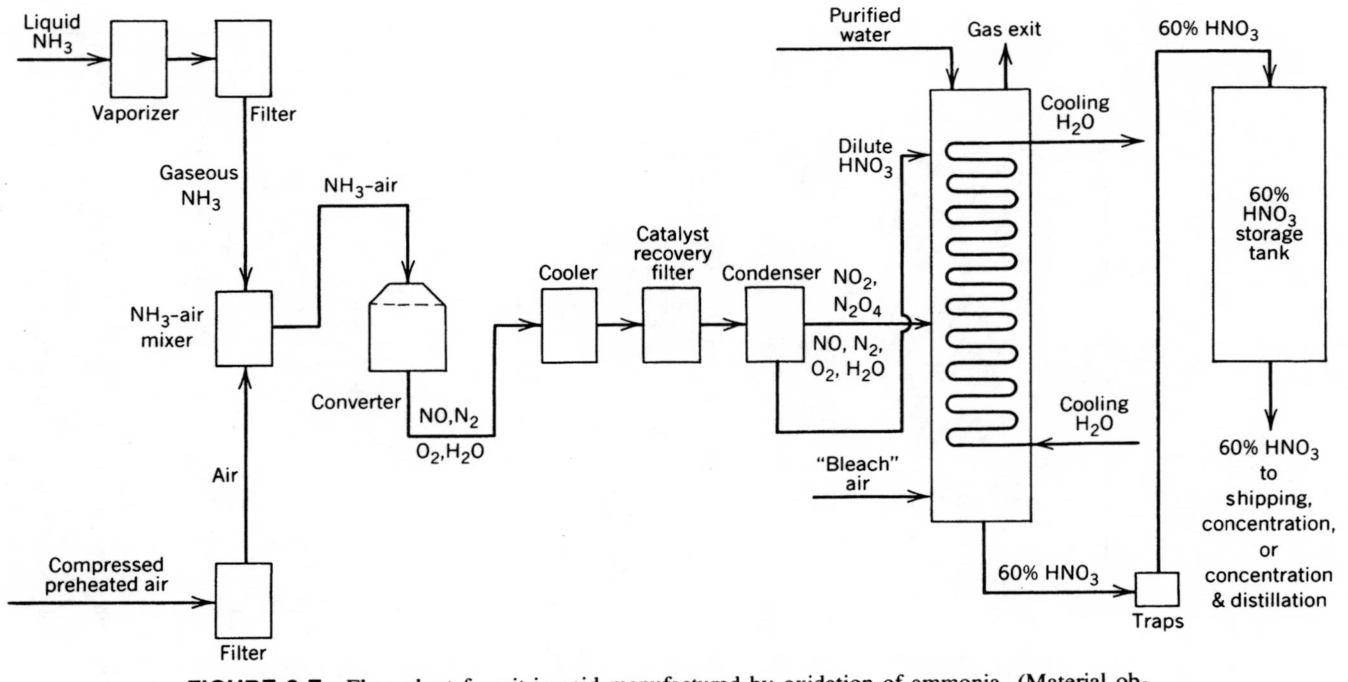

FIGURE 2.7 Flow chart for nitric acid manufactured by oxidation of ammonia. (Material obtained through the courtesy of C-I-L, Inc., North York, Ontario.)

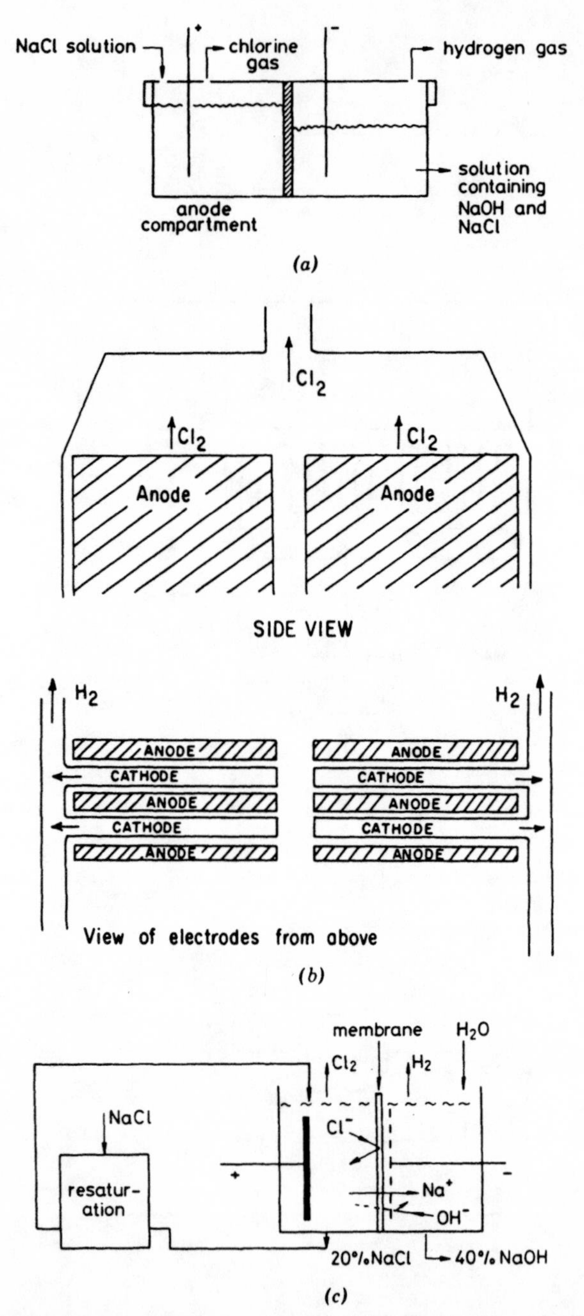

FIGURE 2.8 Simplified electrolysis setup for sodium hydroxide and chlorine. ((*a*) and (*c*) reproduced with permission from R. W. Purcell, *The Modern Inorganic Chemicals Industry*, R. Thompson, Ed., Copyright Royal Society of Chemistry. 1977, pp. 110, 121. (*b*) From Cook, *Survey of Modern Industrial Chemistry*. Copyright 1975 Ann Arbor Science. Used with permission of the publisher.)

or a membrane to prevent the hydrogen from reacting with the chlorine, and for other reasons as stated below. Hydrogen is liberated at the cathode, and chlorine at the anode.

$$2e + 2H^+ + 2OH^- \rightarrow H_2 + 2OH^-$$

$$2Cl^- \rightarrow Cl_2 + 2e$$

2.6.1 Diaphragm Process

The diaphragm is a layer of asbestos on a framework. Figure 2.8*b* shows how the H_2 and the Cl_2 are kept separate. These reactions proceed until about 50% of the sodium chloride is used up. The overall reaction is

$$2Na^+ + 2Cl^- + 2H_2O \rightarrow 2Na^+ + 2OH^- + H_2 + Cl_2$$

The diaphragm extends into the electrolyte solution. The brine is fed into the anode side. The level of liquid in the anode is kept higher than in the cathode to force the liquid brine into the cathode and depress the backward flow of hydroxyl ions. If this happens, reactions such as (1)–(5) will occur:

$$Cl_2 + OH^- \rightarrow Cl^- + HOCl \quad (1)$$

$$HOCl \rightarrow H^+ + OCl^- \quad (2)$$

$$2HOCl + OCl^- \rightarrow ClO_3^- + 2Cl^- + 2H^+ \quad (3)$$

$$4OH^- \rightarrow O_2 + 2H_2O + 4e \quad (4)$$

$$4OH^- + C \rightarrow CO_2 + 2H_2O + 4e \quad (5)$$

With reactions (1)–(3) there will be a loss of chlorine and sodium hydroxide. Reaction (4) results in the formation of oxygen, and this is the routine analysis that is done to see that the electrolytic cell is working correctly.

Reactions (1)–(5) take over when the electrolysis is continued beyond 60–65% of the brine charged. This is why the commercial electrolysis is only taken to 50% of the brine used. The rest of the salt is recovered by centrifugation of the concentrated NaOH, and is reused in the next batches (see Figure 2.9)

A. *The Electrodes*

We now consider the electrodes used in the large scale cells. Anodes used to be made of graphite. Equation (5) indicates how this would become corroded. The distance between anode and cathode would increase and more current would be required. The anodes are now "dimensionally stable anodes" made of titanium covered with ruthenium oxide or palladium oxide. Chlorine does not attack them. The material is called DSA.

The cathodes are coarse steel mesh coated with asbestos to make the diaphragm. They are shaped to keep the H_2 and Cl_2 separate.

B. Manufacturing Process Details

As the following procedure is read, Figure 2.9 should be studied.

1. Brine purification is always necessary to lessen clogging of the diaphragm and to make a purer NaOH. This is done by addition of sodium carbonate and sodium hydroxide for removal of Ca, Mg, and Fe carbonates which are removed from the tank by settling or filtration. The solution is then neutralized with HCl.

2. The brine electrolysis is carried out at 65–75°C. The brine is fed into the anode compartment to maintain a constant level.

3. The work-up procedure involves getting rid of the unreacted sodium chloride. This is performed by evaporation of the electrolyte to 50% concentration. At this stage the remaining NaCl is almost insoluble, and it is centrifuged and

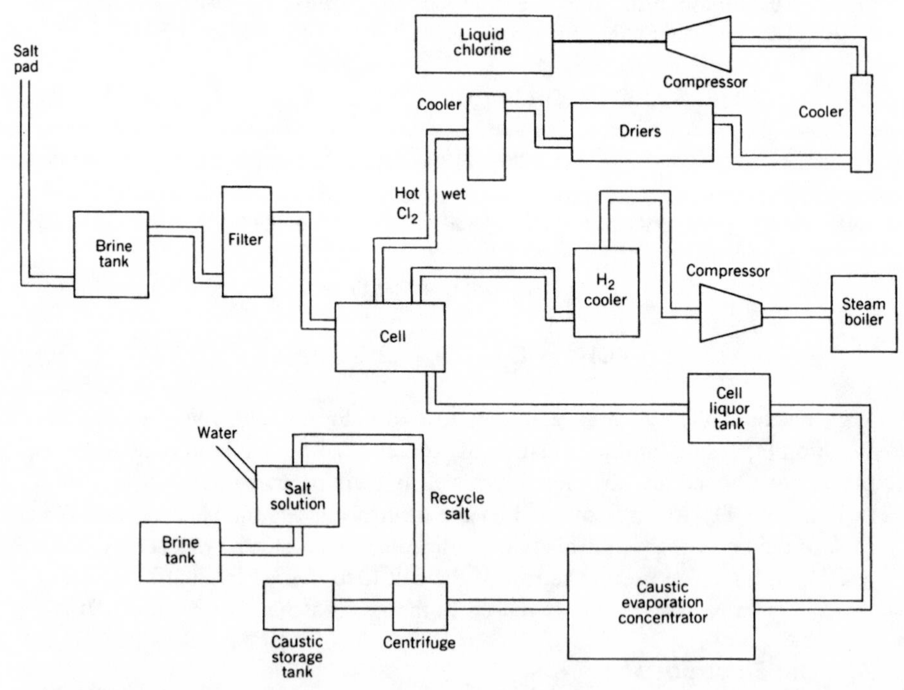

FIGURE 2.9 Flow sheet for a chloralkali diaphragm cell and associated equipment. (From Occidental Petroleum Co. of Canada.)

reused. The solution containing the NaOH is used as is or evaporated to a dry solid in nickel pots. The hydrogen is sometimes used as an energy source by burning on site.

4. The chlorine is dried by passing it through a drying tower having a spray of sulfuric acid and it then is refrigerated and stored as liquid chlorine.

It should be noted that the asbestos membrane has come into disrepute because of the known pollution problem with this material. This has led to a process modification called the *membrane process*. Instead of asbestos we have a Dupont membrane material called Nafion, which permits the passage of Na^+ and OH^- but not Cl^-. This eliminates the concentration step to remove the precipitated brine. It is one of the few materials that can withstand the chlorine–alkaline environment. See Figure 2.8*c*.

2.6.2 Mercury Cell

In the mercury electrolysis cell the anode is DSA and the cathode is a moving stream of mercury. Chlorine gas is formed at the anode and sodium amalgam is formed at the cathode. The equation at the anode is the same as that in Section 2.6. That for the cathode is

$$2Na^+ + 2e + Hg \rightarrow Na_2Hg$$

The amalgam flows to a secondary electrolysis cell called the denuder. Here the amalgam is the anode, the cathode is iron or graphite, and the solution is dilute NaOH. Water is fed in and hydrogen is evolved. The concentration of the NaOH rises to 50%. See Figure 2.10.

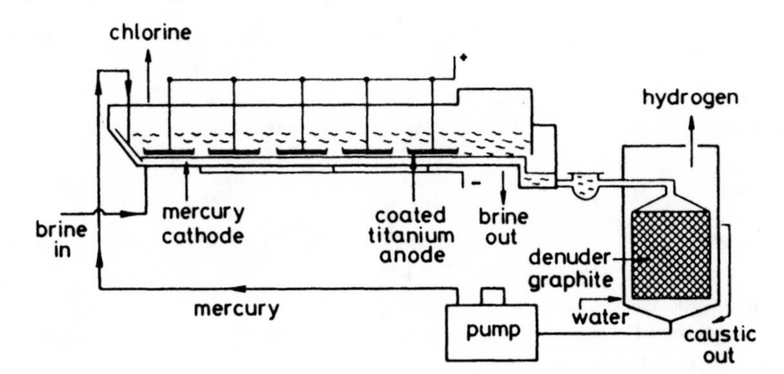

FIGURE 2.10 Sodium hydroxide and chlorine mercury cell setup. (Reproduced with permission from R. W. Purcell, *The Modern Inorganic Chemicals Industry*, R. Thompson, Ed. Copyright Royal Society of Chemistry, 1977, p. 111.)

$$2Na_2Hg \rightarrow 2Na^+ + Hg + 2e \text{ (anode)}$$

$$2H_2O + 2e \rightarrow 2OH^- + H_2 \text{ (cathode)}$$

This process has been the source of much mercury pollution, and no new mercury plants are being built.

An excellent comparison of the three cells was given in *Chemical and Engineering News* (1978). It showed that the membrane cell is superior in forming a more concentrated sodium hydroxide solution, with very low chloride. It greatly reduces the concentration step, eliminates the filtration of NaCl, and eliminates pollution problems from asbestos and mercury. It, however, has a considerably larger power requirement than the diaphragm cell. See Table 2.1.

2.7 SUMMARY

In this chapter we have covered some representative inorganic processes sufficiently to show what sort of problems are encountered that demand chemical know-how. You should now be acquainted with the chemistry of the processes and the effluent problems associated with them. Some of the oldest procedures are still being studied because of cost and environmental considerations. This

TABLE 2.1 Comparison of Characteristics of Three Chloralkali Cells

	Cell Type		
	Membrane	Diaphragm	Mercury
Cell separator	Nafion[a]	Reinforced asbestos	None
Method of separator application	Hanging sheet	Vacuum deposition	—
Anode	DSA[b]	DSA	DSA
Cell caustic			
Wt NaOH	28%	11%	50%
Wt NaCl	50 ppm	15%	30 ppm
Cell power consumption (dc kW per ton NaOH)	2600	2180	2800
Evaporation required (lb H_2O/ton 50% NaOH)	1570	5730	0

[a] Trademark for a DuPont perfluorosulfonic acid polymer.
[b] Dimensionally stable anode, a patented development of Diamond Shamrock.

in the case with sulfuric acid where a special catalytic setup was developed recently to have the yield as high and the effluent and cost as low as possible.

REFERENCES

Andrew, S. P. S., Modern Processes for the Production of Ammonia, Nitric Acid and Ammonium Nitrate, in R. Thompson, Ed., *The Modern Inorganic Chemicals Industry*, The Chemical Society, London, 1977, pp. 201–228.

Bland, W. J., *Educ. Chem.*, **21,** 7–10 (1984).

Chemical and Engineering News, Chloralkali Membrane Cell Set for Market, *56,* March 20 (1978), pp. 21–22.

Childs, A. F., Phosphorus, Phosphoric Acid and Inorganic Phosphates, in R. Thompson, Ed., *The Modern Inorganic Chemicals Industry*, The Chemical Society, London, 1977, pp. 375–392.

da Rosa, A. V., *CHEMTECH* **8,** 28–29 (1978).

E. Emmett, *Educ. Chem.*, **12,** 81–83 (1975).

Kulp, R. L., and D. W. Leyshon, Commercial Processes—Dorr–Oliver, in A. V. Slack, Ed., *Phosphoric Acid*, Vol. 1, Part 1, Dekker, New York, 1968, pp. 213–240.

Phillips, A., The Modern Sulfuric Acid Process, in R. Thompson, Ed., *The Modern Inorganic Chemicals Industry*, The Chemical Society, London, 1977, pp. 183–230.

Purcell, R. W., The Chlor-Alkali Industry, in R. Thompson, Ed., *The Modern Inorganic Chemicals Industry*, The Chemical Society, London, 1977, pp. 106–133.

Shreve, R. L., and J. A. Brink, *Chemical Process Industries*, 4th ed., McGraw-Hill, New York, 1977.

Wittcoff, H., and B. G. Reuben, *Industrial Organic Chemicals in Perspective*, Part 1, Wiley-Interscience, New York, 1980.

EXERCISES

Students should be able to describe the processes with the aid of unlabeled process diagrams.

1. *Phosphoric acid.* Compare the two processes as follows:

Supply equations for conversion and effluent.
Describe the chemical process steps for each vessel in the diagrams.
Compare the purity of the product from the processes.
Why are two processes needed?

2. *Superphosphate.* Explain why triple superphosphate is more concentrated than the normal form.
3. *Ammonia.* Give equations for manufacture from CH_4 and N_2. What catalysts are used?

   Describe the six process steps for each of the vessels in the diagram.

   How is the $CO:H_2$ ratio changed for the syngas prepared for other products?

   What waste products are formed? Can they be used?

4. *Sulfuric acid.* Know the equations for each process step.

   Describe the three process steps and the raw material purification occurring in the vessels in the diagram.

   Discuss choice of raw materials available.

   Discuss the catalyst for the catalytic oxidation. Why is the yield in this reaction so important? How has it been increased in recent years?

5. *Nitric acid.* Know the equations for each step and for the important by-products.

   Describe the three steps occurring in the vessels in the process diagram.

   Know how the by-products are controlled by the catalyst construction.

6. *Sodium hydroxide and chlorine.* Know the equations for the anode and cathode reactions in each of the three processes, and also those for the by-products.

   Understand the diagrams and how the H_2 and Cl_2 are kept apart.

   Know how the raw materials are purified, and the reason for doing so.

   Compare the three processes under the following headings: purity of NaOH, effluent, and power requirements.

3

Products of Fermentation Processes

We are rapidly becoming aware that petroleum supplies, on which 90% of the organic chemical industry depends, are not present in unlimited quantities. So, before considering petrochemicals, we will examine products from various other sources of raw materials. First we will discuss those suitable for conversion to products by fermentation processes, that is, processes depending on the presence of bacteria to convert raw materials to products, as the bacteria population grows.

3.1 BACKGROUND FOR FERMENTATION PROCESSES FOR INDUSTRIAL CHEMICALS AND FOR SEWAGE TREATMENT

We describe in some detail how industrial fermentation reactions are carried out because this is not covered in a typical chemistry major course. Fermentations are used in two distinct ways:

1. Bacterial conversions are of basic importance in sewage treatment. The conversion to innocuous products of waste products that would damage the environment is carried out mainly by a bacterial procedure, as shown in Section 3.2.4.

2. As petroleum stocks shrink, more and more industrial fermentations will be used for making industrial chemicals. Such well-known solvents as ethanol and butanol and such materials as antibiotics are examples of materials that can be made by this process. The aim in a successful industrial chemical production is to make a single chemical product from a few raw materials by the action of microorganisms. Only a single organism is used in industrial chemical fermentations. An introductory review has recently been issued (Kovaly, 1982). It discusses converting readily renewable plant materials (biomass) into chemicals. Some of the procedures discussed involve fermentations. As in all other chemical reactions in industry, high yield and low cost are necessary.

We have the same aim—conversion of raw materials to products in high yield—in water treatment in a sewage treatment plant. In the latter situation, however, conversion is desired of a multitude of chemical waste products to degraded materials by a large number of strains of bacteria. Again high-yield microbiological processes are involved if the sewage treatment is to be successful. We are going to discuss the industrial processes first, and now shall discuss briefly the factors controlling these kinds of processes.

3.1.1 How Organisms Bring About Chemical Conversions

There are two general types of fermentation processes, which differ in their oxygen requirements.

A. Anaerobic Processes

Some bacteria are anaerobic and require no oxygen. The process details depend on the physiological requirements of the particular organism involved. As examples of this type of procedure, fermentation ethanol from grain was recently reviewed by Oliver et al. (1982) for industrial alcohol manufacture and Kent (1983, p. 668) for beer manufacture. The evolved carbon dioxide is vented or used to blanket another vessel to exclude oxygen. We give an example of an anaerobic procedure, the preparation of butanol and acetone from corn, in Section 3.1.5 (Walton and Martin, 1979; Noon, 1982).

B. Aerobic Processes

Aerobic processes require oxygen and are usually carried out in the laboratory in shaker flasks that are gently oscillated to increase oxygen uptake. See Figure 3.1. The microorganisms, requiring oxygen, grow on the surface of the medium containing the chemicals to be converted. Sterile air has to be introduced to prevent extraneous reactions from occurring because of the presence of other bacteria. A sample of the chosen microorganism is then added to convert the materials in the medium. An example of this type of process, the preparation of penicillin, is given in Section 3.1.4.

3.1.2 Prerequisites for a Successful Industrial Fermentation

The typical microorganisms used in industrial fermentations are of four types. The first, bacteria, are one-celled organisms capable of independent growth and multiplying by simple cell fission. Then we have fungi, which exist in environments lower in humidity than bacteria. They have aerobic metabolism and grow in long, filamentous cells called mycelia. Then we have yeasts, which are a sort of fungi but have elliptical cells and grow in number by budding, not cell division. Finally, there are the actinomycetes, which are intermediate between the bacteria and fungi. They are the source of many antibiotics and grow in fermentations similar to fungal fermentations.

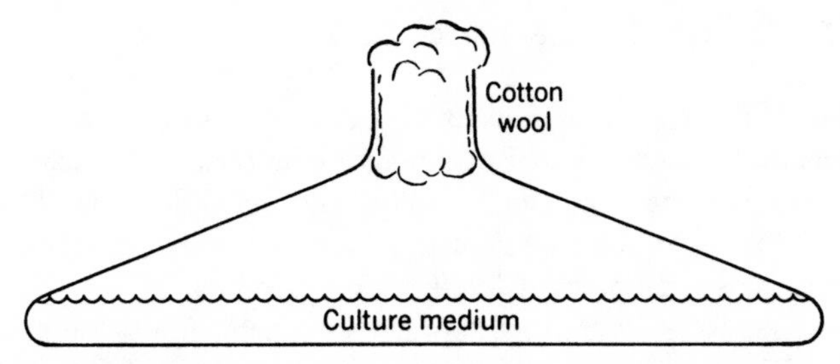

FIGURE 3.1 Diagram of a fermentation shaker flask.

A. *Microbial Nutrition*

It is essential that the organisms stay alive during the process. For food for the organism, we now consider the chemical raw material, the components of which become changed to the products desired or incorporated in the bacteria during the metabolism. In general, the microorganisms need a carbon source, usually a carbohydrate, and a nitrogen-containing material (often a protein), along with trace metals and vitamins. Some microorganisms if supplied with glucose, an inorganic nitrogen source, and some metal salts can synthesize all their amino acids and proteins. Others are unable to prepare essential parts of their systems, and the missing ingredients must be supplied. See the penicillin procedure for background on this (Section 3.1.4).

B. *Source of Carbon for the Cell's Metabolic Systems*

Various carbohydrates and also hydrocarbons can be used as sources of carbon for fermentation processes.

Carbohydrate. A cheap carbohydrate source is necessary in an industrial process because the carbon source is a substantial part of the cost. For a general discussion of process cost see Chapter 6. The following raw materials are examples of cheap carbohydrate sources:

Blackstrap molasses
Beet molasses
Glucose and xylose in sulfite waste liquors
Corn after the oil has been removed

Hydrocarbons. Yeasts and bacteria can be grown to obtain 50% protein using various hydrocarbons as a carbon source. The filtered product is called SCP (single-cell protein). A review of this process is given in articles by Litchfield

(1978, 1979). Kent (1983, p. 684) gives general background on this and other processes where large-scale industrial fermentations are carried out.

The metabolism of microorganisms results in release of energy from food (Emden–Meyerhoff scheme, Krebs cycle) and accumulation of such products as ethanol, butanol, acetone, acetic acid, and fumaric and citric acids. If a particular enzyme is missing, intermediate products accumulate in the preceding schemes. This describes the source of most fermentation products.

When the chemist is trying to determine a fermentation process in order to develop a useful product, the following attributes should prevail. The chemist should know what foods the bacteria need. Also the fermentation should be rapid, the yield should be high, and the workup should be easy. The microorganism must be readily propagated and must maintain biological and biochemical standards to give consistent yields. Bacterial strains must be kept from mutating. Control of by-products has to be maintained; for example, bacterial oxidation often yields acidic by-products that kill the bacteria.

Economically attractive raw materials in good supply and predictably uniform in composition must be available. For this to happen, the medium has to be free of pollutants toxic to the bacteria.

3.1.3 Description of a Fermentation Process

The organism has to be kept pure and uncontaminated in the laboratory. It is propagated on a medium consisting of agar that has been treated with sugars and plant or animal tissue extracts. First, the hot medium is placed in a sterile tube, which is plugged with cotton and sterilized in an autoclave 30 min at 15 psi. After the tube is cooled, the bacteria sample is added at the predetermined temperature. A portion of the growing organism called the inoculum is regularly transferred to fresh sterilized medium. This is to prevent biological changes in the bacteria because of changes in the fermenting medium or buildup of toxic substances. Transfer to a fresh medium prevents the concentration of product from building up beyond a certain point. Table 3.1 shows the wide variation in yield of product, depending on the fermentation being studied.

The inoculum of bacteria is placed in the raw material mixture. The reaction shows the following stages:

1. There is a lag time while the culture is multiplying to maximum cell count.
2. Product fermentation then proceeds at its maximum rate.
3. Next there is declining productivity as the substrate is exhausted and toxic substances accumulate.

After the microorganism has brought about the conversion, the waste bacterial cell material is removed by filtration, and the mother liquor is typically separated

TABLE 3.1 Maximum Concentration in Medium for Various Fermentation Products

Product	Maximum Concentration
Organic acids	150 g/L
Antibiotics	100s of ppm
Vitamins	100s of ppm
Vitamin B-12	1 ppm

into acidic, basic, and neutral fractions, as is usual in chemical reaction workups. For example, organic acids are precipitated as calcium or barium salts, penicillin is partitioned between solvent and buffer, and solvents such as butanol are distilled.

A. *Deep-Tank Fermentations*

Aerobic processes were initially carried out in large shaker flasks (see Figure 3.1), in which the reaction proceeds on the surface of the liquid, oxygen being absorbed from the air. However, for larger quantities, deep tanks of many thousands of gallons capacity, in which air is blown through the mixture, are used. Fermentation of such materials as penicillin is carried out in this way. Conditions have been worked out so that the reactions are completed in a desirable time and have a high yield. Sterile air (one volume of air per volume of medium per minute) has to be blown in to keep the reaction going. The air is sterilized by passing it through a sterile filter packed with an agent such as glass wool. The packing gland around the agitator has to be sterile as well as the antifoam used.

In an article on practical aspects of industrial fermenter design (Knopfel and Muller, 1978), the following points were emphasized concerning reactions in deep tanks:

First, high oxygen transfer is essential in the reactor, so the setup inside the vessel should ensure the following:

1. The surface area of the oxygen in contact with the reaction mixture should be as large as possible. That is, the gas should be present as many very small bubbles.
2. The reactor should be held above atmospheric pressure.
3. Oxygen should be used instead of air.
4. The diffusion coefficient in the system should be as favorable as possible.

The use of antifoams is sometimes necessary because the air blowing through the reaction causes excessive foaming. This may cause unfavorable oxygen absorption conditions.

Second, there should be a zone of intensive mixing. This is of vital importance in maintaining reaction rate.

Finally, many fermentations, for example, SCP (single-cell protein), liberate a lot of heat. The equipment should have an effective cooling system with sufficient supply of low-temperature cooling water.

Figure 3.2 shows a typical fermenter.

B. Change of Fermentation Process to Favor a Desired By-product

In many cases a fermentation process yields one economically desirable substance plus additional by-products. After the passage of time, a certain by-product may become more valuable commercially than the original product. An example of this is vitamin B-12 from the mother liquors of the antibiotic streptomycin. The fermentation conditions were changed by process development work to increase the yield of the former by-product, now the product.

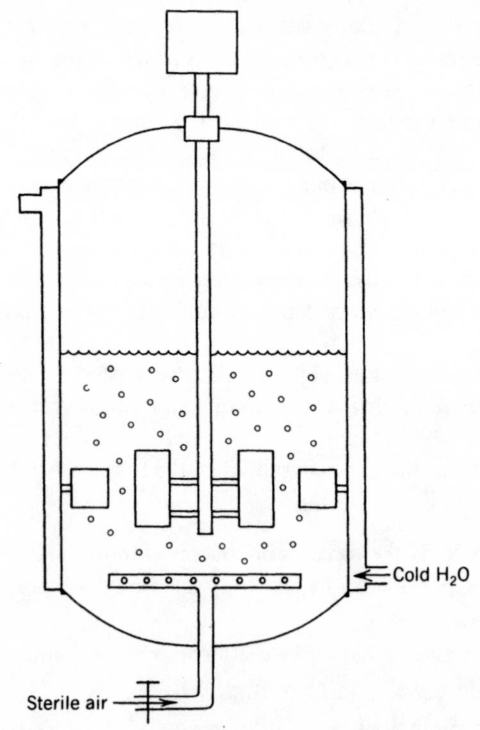

FIGURE 3.2 Diagram of a typical fermenter. (Adapted from Knopfel and Muller, *Chem. Ind.* (London), 782 (1978).)

3.2 INDUSTRIAL FERMENTATIONS USING AEROBIC AND ANAEROBIC METHODS

3.2.1 Penicillin G—An Aerobic Fermentation

(Penicillin G *or* 6-(phenylacetamido) penicillin.

We are now going to study penicillin manufacture as an example of the various general points made above. A review of this work is given in Perlman (1979, p. 255–259). This is an aerobic process. The modern process was formulated by changing the initially used *Penicillium notatum* and simple chemicals supplemented with such substances as casein digest, yeast extract, or meat products. A much better yield was obtained by using corn steep liquor and changing to P. *chrysogenum* as the source organism. The yield of antibiotic greatly improved, because the corn steep liquor contained what are now known as metabolic precursors to the antibiotic, phenylalanine and phenethylamine. The next step was to define chemically what was needed in this process. Table 3.2 shows

TABLE 3.2 Comparison of Improved Medium with Corn Steep Liquor Medium[a] in Penicillin Manufacture

	Percent of Medium	
Type of Ingredient	Corn Steep Liquor	Improved Medium
Main carbohydrate	Lactose, 3.0–4.0	Lactose, 3.0
Other carbohydrate	Glucose, 0–0.5; polysaccharides various concentrations	Glucose, 1.0; starch 1.5
Organic acids	Acetic, ~0.05 Lactic ~0.5	Acetic, 0.25; citric 1.0
Special precursors	Phenethylamine[b]	Phenylacetic acid, 0.05
Main N source	Amino acids, amines, peptides	$(NH_4)_2SO_4$, 0.5
Other N source	Ammonia	Ethylamine, 0.3
Total solids	8.0–9.0	8.5
Total N	0.12–0.2	0.2

[a] See Perlman (1979).

[b] Phenethylamine, found in many batches of corn steep liquor, was often increased in these media by addition of pure phenethylamine or phenylacetic acid supplements.

the ingredients in the penicillin medium, and it should be studied along with the following comments. Perlman uses this table as a basis for showing the origin of the improved more chemically defined medium that followed from the corn steep liquor.

The following improvements were made in the penicillin process to improve yield and reduce cost:

1. Replacing expensive lactose with continuous addition of glucose or sucrose.
2. Controlling pH of the fermentation to keep it between 6.8 and 7.4, using buffers ($CaCO_3$ and phosphates).
3. Controlling pH by the use of glass electrodes. These had to be able to withstand repeated autoclavings.
4. Maintaining high levels of aeration and agitation.
5. Maintaining temperature at 25° ± 0.5°.

A. *Penicillin Production by Fermentation*

We shall now briefly discuss the penicillin production process, using the flow diagram in Figure 3.3

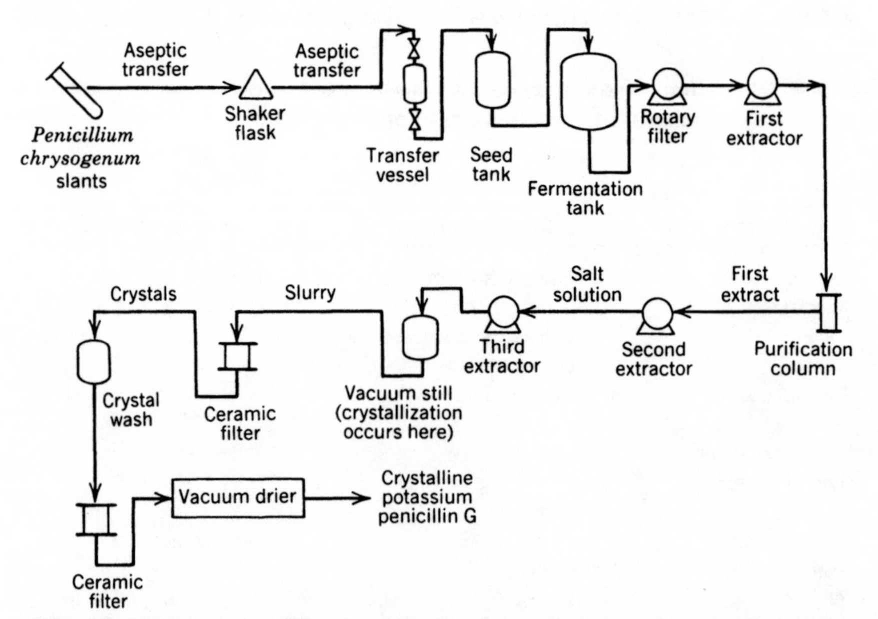

FIGURE 3.3 Flow sheet for penicillin production. (Reproduced with permission from D. Perlman, *Microbial Technology*, 2nd ed., H. J. Peppler and D. Perlman, Eds. Academic Press, New York, 1979, Vol. 1, p. 259.)

1. Spores of P. *chrysogenum* are used to inoculate 100 mL of medium in a 500-mL Erlenmeyer flask.
2. The flask is incubated for 4 days on a shaker at 25°C.
3. The contents of the flask are transferred to a 4-L flask containing 2 L of medium and incubated 2 days at 25°.
4. The contents of the flask are transferred to 500 L of medium in an 800-L stainless steel tank, equipped as in Figure 3.2 and incubated for 3 days.
5. The contents of the reactor are transferred to 180,000 L medium in a 250,000-L fermenter and incubated for 5–6 days.
6. The contents of the vessel are filtered. The filtrate is passed through a series of Podbielniak extractors where the penicillin is extracted into amyl acetate.
7. The penicillin is extracted into aqueous phosphate buffer from the amyl acetate.
8. The potassium penicillin G is crystallized from a butanol–water mixture or converted to the procaine salt.

3.2.2 Butanol–Acetone—An Anaerobic Fermentation

For an interesting account of the history of this process, see Walton and Martin (1979, pp. 188–189). This process is an anaerobic one. The organism in this process (see Walton and Martin, 1979) is called *Clostridium acetobutylicum*. The raw materials are corn, blackstrap molasses, high-test molasses, or other carbohydrates. As stated earlier, the carbohydrate is the largest cost factor. Waste carbohydrates such as hydrolyzed wood, hydrolyzed corncobs, and other sorts of materials should be chosen.

We examine the procedure using corn. First, to remove the corn oil, a valuable commercial product, a steeping process is carried out. The remaining mash is added to 60% water and 40% stillage (fermented liquor from which the solvents have been distilled). The tanks have been previously sterilized with steam. A small amount of superphosphate is added and also some ammonia for pH control at 5.5–6.0. The fermentation equipment is heated 4–6 hr at 110–115° and then cooled, while it is maintained under a pressure of 15 psig sterile fermenter gas (carbon dioxide and hydrogen). This ensures an anaerobic sterile environment.

The details of the preparation of the spore-forming organism are given in Walton and Martin (1979, pp. 194–195). The fermentation is carried out first in a 1000-gal tank and then this mixture is added to a 50,000-gal fermenter. Steam is added to give a slight positive pressure. In 50–60 hr the process is complete. Product: an aqueous solution containing acetone, butanol and ethanol. Yield = 0.25–0.26 lb. solvents per lb dry corn. The concentration of total solvents in water is 22.3 g/L. The solvents are present in a ratio of 30% acetone/60% butanol/10% ethanol.

The solvents are separated by a continuous fractional distillation procedure.

3.2.3 Cost of the Fermentation Approach

An interesting example of low-cost fermentation has been described by Brierly, (1982). Recovery of copper concentrates from low-grade wastes by a fermentation technique is described. Bacteria convert insoluble copper sulfide into water-soluble copper sulfate, readily leached by water. This process is used to supply up to 10% of the copper used in the United States.

A very low cost process is the production of vinegar (a water solution of acetic acid). This product is made from 10% ethanol containing phosphates and other inorganic salts. This solution is passed through beechwood shavings containing *Acetobacter aceti*, moistened with acid. The yield is high and the sale price low.

An example of a high-cost process is vitamin B-12. The yield is low, but, considering that the material is active in a dose in the microgram range, a low yield is commercially feasible. The sale price of this product per pound is extremely high.

With a high yield process the main factor against a low cost is that most fermentations are very dilute processes. The concentration in kilograms per batch is usually very low. The isolation step can be expected to be a significant cost factor.

A. Factors Influencing Fermentation Yield

1. The environment for microorganisms should be optimum. The substrate ingredients have to have good nutritional value. Many microorganisms produce acids, and pH control is necessary. If acids build up, the rate of fermentation falls off. Basic substances are added for pH control, for example, calcium carbonate.
2. Temperature control usually has to be within 1–2°C. A high temperature destroys culture, and a lower temperature depresses rate of fermentation.
3. Adequate aeration and mixing is vital.
4. As low a temperature and as short a time of fermentation as possible should be kept; otherwise toxic substances are formed and nutrients destroyed.
5. The size of the bacterial inoculum varies with each process, and this is found by trial and error.

3.2.4 Comparison of Industrial Fermentations and Sewage Treatments

Before going on to water treatment and sewage treatment, we shall now establish the connection between the preceding data on industrial fermentation and what occurs in natural water systems. There the bacterial processes act as decomposers of dead plant and animal life to chemical substances usable in the growth of the next generation. When a lake or pond is used as a receptacle for sewage and chemical waste, this natural system is interfered with. So we have to consider how to convert polluting substances to innocuous nontoxic products. This is carried out by degrading these substances by microbiological processes in the waste treatment plant.

For an industrial process one specific organism is required. In sewage systems many species of bacteria are needed for the degradation of many substances. In other ways the two processes are similar, and the necessary conversions of sewage components have to be studied in the same manner as industrial fermentations. In both cases we require fast conversion, high yield, and the development of conditions for complete conversion of the material.

In sewage treatment, bacterial degradation processes are sometimes not sufficient for complete conversions, and chemical treatment has to be done as an additional step.

3.3 WATER—AN IMPORTANT RAW MATERIAL FOR THE CHEMICAL INDUSTRY

Water is essential for humans and animals to drink and for plants to grow. The chemical industry uses it chiefly as a cooling medium, and also as a process solvent and in steam boilers for energy and heat.

3.3.1 Water Uses and Corresponding Analytical Values

We now discuss the five major uses of water. Water for differing uses has varying quality requirements. Water unsuitable for one use may be satisfactory for another. Water in nature has various amounts of dissolved and suspended inorganic and organic compounds in it. The actions of man, particularly in industry, add to this load of dissolved material. Considering this fact we need to decide what sort of water we need for our projected uses. Table 3.3 lists the analytical requirements for water for various uses.

Municipalities and industry use water as a medium to carry away effluent. It is necessary to carefully control the impurity level in the stream or lake receiving

TABLE 3.3 Typical Analyses of Water for Various Uses[a]

Agricultural Uses

Livestock

  low bacteria (<40/100 mL) and low toxic substances

Irrigation

  low dissolved solids (<500 mg/L) to avoid increasing soil salinity, total bacteria; allow 100,000/100 mL

  desirable <10,000/100 mL

Fish, aquatic life, wildlife

  Concentrations of toxic substances low

  pH near neutral, 6.5–8.5

  Low BOD, 1–2 mg/L or less

  Dissolved oxygen: cold, 6–7 mg/L

  warm, 4–5 mg/L

  Low temperature, turbidity, etc.

Industrial Uses

Cooling vessels; steam boilers

  Low hardness <50 ppm Mg^{2+} + Ca^{2+} usually as SO_4^{2-} and CO_3^{2-}

  Food processing, brewing, soft drinks

  As public drinking water, but lower fluoride

Public Recreational Uses

  Free of color, odor, taste, turbidity

  Total bacteria < 1000/100 mL

  Coliforms <100/100 mL low nutrients, to avoid nuisance algae growths

Public Drinking Water (treated)

  No bacteria

  Low nitrates, nitrites <10 ppm

  Very low pesticides <0.05 ppm

  Fluoride allowed to 2.4 ppm

[a] From Dr. M. Hocking, University of Victoria, Victoria, BC, Canada

the effluent, as eventually the water may well be going into the water table over a wide area. Effluent from the chemical industry is composed of many process by-products whose toxicity has not been carefully tested, or not tested at all. This effluent has to be rendered nondamaging to the environment; otherwise release of the effluent should not be approved by the authorities.

Most people know that municipal waste treatment systems convert waste of many kinds into forms nondamaging to the environment. The essential way in which this happens is through the action of bacteria and oxygen. The question with chemical process waste is whether the bacteria in the local waste treatment system will actually convert these by-products. A serious problem could arise if inadequately treated waste water were sent directly to a river or lake having the usual plants, fish, and nonpathogenic bacteria present. There are two situations that could result. First, waste chemical compounds could be rapidly oxidized by bacteria, depleting the oxygen supply in the water below the value essential for the fish and plants to live (see below). Second, these materials (e.g., herbicide waste or heavy metals) could kill fish and plants by a toxic reaction.

A. Measurement of Pollution in a Lake or Stream

A municipal waste treatment plant is used to prevent or greatly reduce pollution of a stream because of added sewage effluent.

The most common index of pollution of a lake or stream is called BOD, biochemical oxygen demand. This figure gives the rate at which oxygen will be used up by the action of bacteria on readily oxidizable effluents (e.g., municipal sewage). BOD is determined by measuring dissolved oxygen (DO), before and after test incubation covering 5 days (for the method, see Section 3.3.2.C).

A nonpolluted body of water can be described as follows:

1. The DO content of the water has to be greater than 5 ppm to permit a satisfactory warm-water biota to exist in it.
2. The BOD has to be low. The BOD of pure water is less than 1 ppm. It is a cause for concern when an effluent stream entering a lake has a BOD of greater than 20 ppm.
3. No excessive nitrogen or phosphorus compounds should be present. The nitrogen and phosphorus level should be limited to the natural level to prevent rapid algal formation. With excessive nutrients an algal "bloom" will occur. Once excess nutrients have been used up, the algae will die and decomposition of waste will deplete the oxygen.
4. No refractory organics (pesticides or herbicides) should be present. These would kill off plants and insects.
5. No pathogenic bacteria should be present if the stream is to be used for drinking.

3.3.2 Discussion of Pollution in Natural Waters

A stream or lake is considered polluted if a sample contains less than 5 ppm oxygen. A BOD of greater than 20 mg/L, a high concentration of nutrients (ammonium or phosphate), and a high concentration of refractory organics also indicate pollution.

The presence of dissolved oxygen is a fundamental requirement for the plant, animal, and bacterial population in any body of water. Fish require the highest level of DO, invertebrates a lower level, and bacteria the smallest quantity. For a diversified warm-water biota, the DO should be greater than 5 ppm.

A. Cause of Water Deoxygenation

The presence of oxygen-demanding wastes, that is, substances easily broken down or decayed by bacterial activity in the presence of oxygen, causes available oxygen to be consumed. Examples of oxygen-demanding wastes are such materials as wastes from humans and animals, food processing, paper mills, tanneries, etc. The carbon in these wastes is oxidized to CO_2 by bacteria.

$$C + O_2 \rightarrow CO_2$$

It takes 32 g of O_2 to oxidize 12 g of carbon wastes.

B. Estimation of Oxygen-Demanding Wastes

There are three tests, commonly called BOD, COD, and TOC, that are available for estimating the presence of oxygen-demanding wastes:

1. BOD (biochemical oxidation demand). The amount of oxygen in milligrams taken up by 1 L of the sample during 5 days incubation of the sample with active bacteria.
2. COD (chemical oxidation demand). The amount of potassium dichromate, expressed as mg O_2/L taken up by 1 L of the sample when treated with hot acidic dichromate.
3. TOC (total organic carbon). The amount of carbon (in mg C/L sample) as shown by the amount of CO_2 evolved in a high-temperature catalytic oxidation

The BOD test does include all sorts of material, particularly from a biological source (e.g. all sorts of human and animal waste and food processing wastes). Some organics (e.g., herbicides) designed for stability against bacterial action are not degraded. The test results include ferrous salts and some nitrogen, phosphorus, and sulfur compounds. The method is subject to errors caused by many variables.

C. Biochemical Oxidation Demand (BOD)

In this test method (Vesilind, 1983) a sample of water to be tested is incubated in the dark for 5 days at 20°C. The dissolved oxygen is determined before and after incubation. The difference in the DO, expressed as mg/L, is the BOD. It is necessary to analyze the DO once a day and plot the results as in Figure 3.4.

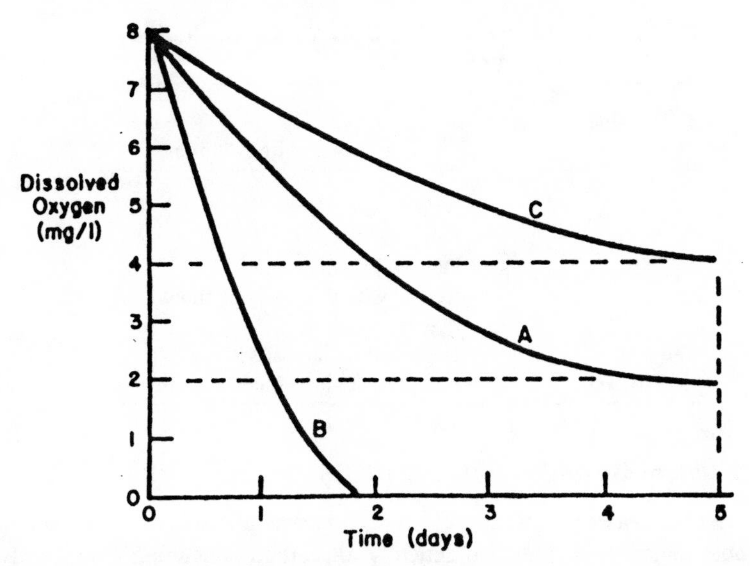

FIGURE 3.4 Graph of BOD tests, showing oxygen uptake. (From Vesilind and Peirce, *Environmental Pollution and Control*, Second Edition, p. 41. Copyright 1983, Ann Arbor Science. Used with permission of the publisher.)

Example 3.1. The graph in Figure 3.4 shows the test results on three samples of river water A, B, and C. The BOD results are:

Sample A: $8 - 2 = 6$ mg/L.

Sample B: DO has dropped to 0 mg/L in 2 days. In this case the BOD value is clearly greater than 8 mg/L because more oxygen would probably have been used if it were present.

Sample C: This sample was prepared by dilution of sample B by 1:10

BOD results:

Sample C: $8 - 4 = 4/0.1 = 40$ mg/L

The obtaining of a low BOD value can have several meanings:

1. The sample of water is fairly pure.
2. The water may still contain certain organic substances, for example, herbicides, which are not oxidizable by bacteria.
3. The microorganisms are dying from pollution by antibacterial substances.

Table 3.4 shows the various BOD values corresponding to pollution from a variety of sources. There are other methods for determining the extent of pollution in a body of water.

TABLE 3.4 Representative BOD Values for Various Pollution Levels

BOD Value (ppm)	Amount of Pollution
1	Nearly pure water
3	Fairly pure water
5	Doubtfully pure water
20	Runoff with this analysis should not enter a stream
100–400	Untreated municipal sewage
100–10,000	Food processing effluent, cattle feedlots

D. Chemical Oxidation Demand (COD)

In this test the water sample is treated with excess potassium dichromate. The oxidizable material, including practically all carbon-containing compounds, is determined by titrating the remaining oxidizing agent.

The COD value is usually higher than the corresponding BOD figure, because many substances not converted by bacteria are oxidized. The test is finished in three hours. It does not detect acetic acid. It will analyze for chloride ion, which is not really a pollutant.

E. Total Organic Carbon (TOC)

This test consists of a complete combustion of carbon in the sample. All carbon-containing substances are oxidized to carbon dioxide. Obviously, noncarbon containing substances oxidizable by bacteria are not included.

The following methods of analysis are used to test for dissolved oxygen (Vesilind 1983).

1. A test meter having an oxygen probe is immersed in the sample and an analysis value is read on the meter.
2. In the Winkler method a manganous salt is added to the oxygen-containing sample. The amount of MnO_2 formed, as determined by an iodometric method, gives the oxygen content.

3.3.3 Basic Purposes of a Municipal Waste Treatment Plant

1. The waste stream enters a system of tanks containing bacteria and remains until the bacterial action is complete. Chlorination is used at the last stage to kill pathogenic bacteria.

2. Toxic nonbiodegradable substances are removed by chemical treatment or absorption, for example, on carbon. The water then can be released to a river or lake.

3.3.4 Removal of Contaminants in Sewage

This section describes the most common methods of municipal sewage treatment. The steps followed in waste treatment plants are quite similar even though the equipment used in various municipalities may appear different.

A. Primary Treatment

Separation by filtration and/or sedimentation of all water-insoluble substances is the first step. The solid is called sludge. It contains raw water solids and is well over 90% water.

B. Secondary Treatment

The next step involves aerobic digestion. Bacteria are allowed to oxidize all biodegradable substances. Air is bubbled in. The growing bacteria utilize some carbon, hydrogen, and nitrogen-containing pollutants in making their own protein and convert the rest to CO_2 and NH_4^+. The sludge, which contains dead bacteria, is separated.

There are two pieces of equipment used to carry out the preceding operation

1. *Trickling filter:* The effluent from the primary process is passed through a large bed of broken stones or slag. Bacteria growing on the rocks react with the waste substances in the water. The bacteria thickness is controlled by grazing insects and protozoans.
2. *Activated sludge:* The waste water is aerated to supply oxygen to the growing organisms (see Fig. 3.5). The sludge is removed by filtration. In this

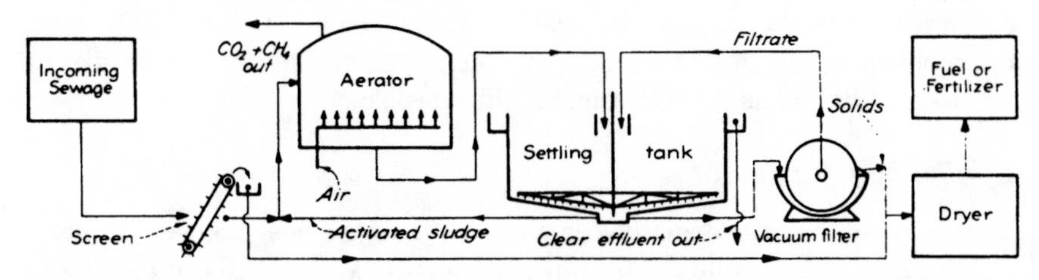

FIGURE 3.5 Diagram of a municipal waste treatment aeration system. (Reproduced with permission from R. N. Shreve and J. Brink, *Chemical Process Industries*, 4th ed., copyright 1977, McGraw-Hill, New York, p. 40.)

case, however, the material consists of dead bacteria. Some of the aerated mixture is returned to the main chamber to act as inoculum.

The treatment of water to remove polluting substances basically achieves the conversion of these materials to solid sludge. If left in the water, they would be dispersed all over the water table. The large quantity of sludge has become an ever-increasing disposal problem as secondary treatment becomes more widespread.

C. Sludge Handling and Disposal

The difficult job of sludge disposal has four main objectives:

To convert the solid material, which consists of unstable, odorous substances, to more stable forms.

To remove liquids.

To destroy or control pathogenic organisms.

To obtain substances that can be used or sold, since the cost of sludge handling is high.

To reach these goals, the following steps are carried out, as reviewed in Christman (1978) and Vesilind (1983):

Sludge Thickening. The sludge is caused to settle, often by the addition of thickening agents such as alum or flotation agents that act by causing the solids to rise to the liquid surface after aeration.

Stabilization. A further digestion step is carried out under anaerobic (with some aerobic) fermentation. Solids are decomposed, the material becomes less odorous, and pathogens are reduced. Methane, useful as a fuel and in many other ways (see Section 4.8) is formed in significant quantity. Other treatments are discussed in Christman (1978) and Vesilind (1983). After drying or combustion it is ready for final disposal. The treated material can be used as landfill (there is a problem here if, for example, heavy metals are present) or as a fertilizer if the material has an analysis that makes this feasible.

D. Tertiary Treatment

The aim in tertiary treatment is to remove all the nonbiodegradable material.

Nitrogen can be eliminated by adjusting the pH to around 10 and by blowing air through the system to remove ammonia gas.

$$NH_4^+ + OH^- \rightarrow NH_3 + H_2O$$

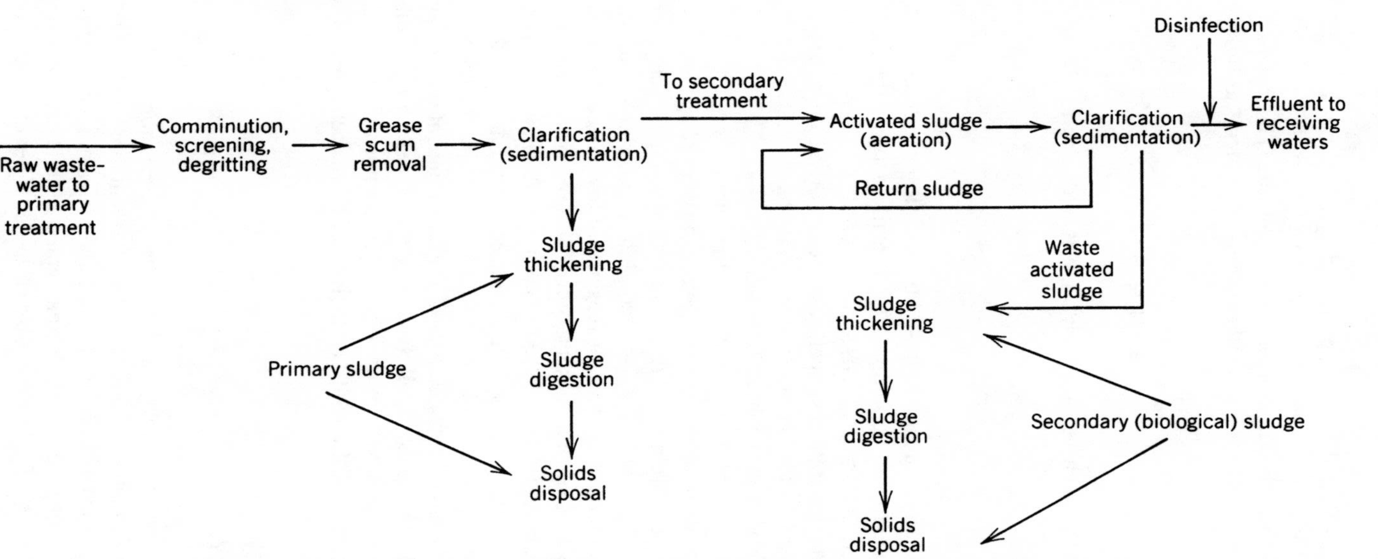

FIGURE 3.6 Basic primary and secondary treatment scheme for municipal waste treatment system. (Reprinted with permission from R. F. Christman, *Cleaning Our Environment—A Chemical Perspective*. Copyright 1978 American Chemical Society, p. 215.)

Removal of nitrogen can also be acomplished by the nitrification–denitrification process. This process is performed using two types of bacteria. One strain in the first tank oxidizes the ammonium ion to nitrate. The reaction mixture is reduced in the second tank by a another strain of bacteria to nitrogen gas that is blown off.

$$NH_4^+ \xrightarrow{\text{bacteria}} NO_3^- \xrightarrow{\text{bacteria}} N_2$$

For removal of phosphate, ferric chloride is added to as shown by precipitation

$$PO_3^- \xrightarrow{FeCl_3} FePO_4$$

It can be added as an extra step followed by filtration or added to the secondary treatment and filtered with the sludge. This is really only acceptable if the sludge is to be incinerated. Alum or $Ca(OH)_2$ can be used as a phosphate precipitatior.

Toxic nonbiodegradable chemicals can be absorbed on a bed of charcoal. Examples of these include the many chemicals that have been designed to be stable in the environment, such as herbicides.

Pathogenic bacteria are removed by chlorination or ozone treatment just prior to release of the effluent to the river or lake. Cook (1982, p. 392) discusses the use of ozone. A new, relatively nontoxic treatment agent has been proposed (Worley, 1983). It is 3-chloro-4,4-dimethyl-2-oxazolidinone and is stable and active against a variety of bacteria at reasonably low concentrations.

The technique of reverse osmosis to purify water has recently been reviewed (Arden, 1977, Dale and Okos, 1983). This method, initially developed to remove salt from sea water, is used in such areas as Bermuda and Near Eastern deserts, where the only source of fresh water is from rain. Recent war conditions in the Persian Gulf have produced a need for development of a reverse osmosis process that will also remove oil from sea water near broken oil-producing facilities. This has recently been developed (O'Connor, 1982).

Table 3.5 summarizes the relative efficiencies of various types of water treatment.

3.4 SUMMARY

This chapter reviews the background for the fermentation technique used for preparing certain industrial chemicals. The fermentation route often uses raw materials which are not petroleum-derived and are significant alternate sources for these products. In other cases, for example, antibiotics, the use of classical syntheses would not be feasible because of the complexity of the method. The

TABLE 3.5 Removal Efficiencies of Various Sewage Treatment Processes (Percent Removal of BOD in Effluent for Various Combinations of Treatment Methods)[a]

			Primary + Secondary + Tertiary		
Test	Primary	Primary + Secondary	Chem Coag	Shallow Lagoon	R. Osmosis +Elctrdialysis
BOD	35	90	95	95	99
COD	30	80	85	90	99
Refractory organics	20	60	80	85	99
Suspended solids	60	90	95	95	99
Total N	20	50	60	85	99
Total P	10	30	85	85	99
Dissolved minerals	1–2	5	10	10	99

Bacteria removal, 90
After disinfection with chlorine, 99%

[a]From Dr. M. Hocking, University of Victoria, Victoria, BC, Canada.

fermentation approach is also used in commercial waste water treatment, where conversion of waste products to innocuous materials is achieved. A general background in waste water treatment is given.

REFERENCES

Arden, T. V., Water Purification and Recycling, in R. Thompson, Ed., *The Modern Inorganic Chemicals Industry*, The Chemical Society, London, 1977, pp. 95–97.

Brierly, C. L. *Scientific American* **247,** 44 (1982).

Cook, G. A., *J. Chem. Educ.* **59,** 375–402 (1982).

Christman, R., The Water Environment, in T. E. Larson, Ed., *Cleaning Our Environment*, American Chemical Society, 1978, pp. 188–274.

Dale, M. C., and M. R. Okos, *Ind. Eng. Chem.- Prod. Res. Dev.* **22,** 452–455 (1983).

Kent, J. A., *Riegels Handbook of Industrial Chemistry*, 8th ed., Van Nostrand, New York, 1983.

Kovaly, K. A., *CHEMTECH* **12,** 486–489 (1982).

Knopfel, H. P., and Muller M., *Chem. Ind.* (London) 782–784 (1978).

Litchfield, J. A., *CHEMTECH* **8,** 218–223 (1978)

Litchfield, J. A., Production of Single-Cell Protein for Use as Food or Feed, in H. J. Peppler and D. Perlman, Eds., *Microbial Technology*, 2nd ed., Vol. 1, Academic, Orlando, 1979, pp. 93–146.

Noon, R., *CHEMTECH* **12,** 681–683 (1982).

O'Connor, R., U.S. Patent 4,366,063 (1982); *CA* **98,** 110302 (1983).

Oliver, W. R., R. J. Kempton, and H. A. Conner, *J. Chem. Educ.* **59,** 49 (1982).

Perlman, D., Microbial Production of Antibiotics, in H. J. Peppler and D. Perlman, *Microbial Technology*, Academic, New York, 1979, pp. 241–276.

Vesilind, P. A. and J. J. Peirce, *Environmental Pollution and Control*, 2nd ed., Butterworths, Stoneham, 1983.

Walton, M. T., and J. L. Martin, Production of Butanol–Acetone by Fermentation, in H. J. Peppler and D. Perlman, *Microbial Technology*, 2nd ed., Academic, New York, 1979, pp. 187–209.

Worley, S. D., W. B. Wheatley, H. H. Kohl, H. D. Bushett, and J. A. van Hoose, *Ind. Eng. Chem.-Prod. Res. Dev.* **22,** 716–718 (1983).

EXERCISES

1. Discuss the ways in which industrial fermentations are similar to bacterial sewage treatment procedures.
2. List the prerequisites for a successful industrial fermentation process.
3. Define anaerobic and aerobic fermentations.
4. Describe a fermentation procedure. Discuss large-scale (deep tank) procedures, listing two essential points.
5. Using the flow sheet, describe the penicillin G process.
6. Discuss what affects the cost of a fermentation process and give processes that use inexpensive raw materials.
7. Discuss water pollution. Define polluted and nonpolluted water. What is the most important criterion in judging whether water is polluted?
8. Describe methods of measuring pollution—BOD, COD, TOC. Discuss the BOD procedure in detail.
9. Discuss a municipal waste treatment plant, with primary, secondary, and tertiary procedures.

4

Organic Chemical Processes

We now move on to consider processes other than those of the inorganic and fermentation types, in this chapter focusing on the organic chemicals. One of the important goals of this book is to make you aware of the limitations imposed on the producer of chemical products by the availability of raw materials. Natural gas and petroleum are used in the manufacture of seven basic organic compounds and in various synthesis gas compositions, as discussed below. Most of the millions of known organic compounds are derived from these seven major organic compounds, which yield 90% of all organic substances. The seven compounds are ethylene, propylene, butene (the C-4 fraction; see section 4.4), benzene, toluene, xylene, and methane. The synthesis gas process presented in Section 2.3.1 is also the basis of a great deal of organic chemisty in addition to being the source of hydrogen in ammonia synthesis.

4.1 CONVERSION OF PETROLEUM INTO PURIFIED CHEMICAL SUBSTANCES

For general background on this broad topic, the review by Harris and Tishler (1973, pp. 280–297) is recommended. There are two classes of methods to achieve conversion of petroleum into purified substances:

1. Separation methods that do not change the chemical nature of the molecules involved.
2. Conversion processes in which the chemical nature of the molecules is altered.

4.1.1 Separation Processes

A mixture of molecules is separated into two or more streams or cuts, using the following characteristics:

Size: Smaller molecules are separated from larger ones.

Linearity: Molecules with a linear structure are separated from branched ones.

Aromaticity: Aromatic molecules are separated from linear ones.

A. Solvent Extraction

In the laboratory a separatory funnel is used to separate substances by differential solubility in immiscible liquids. In the petroleum refinery two examples of this extraction method are the following:

1. Furfural is used as a solvent to remove color bodies containing sulfur and oxygen.
2. Diethylene glycol is used as a solvent to extract the benzene–toluene–xylene (BTX) aromatic fraction.

B. Absorption

Purification of compounds is achieved by selective absorption of ingredients on alumina or carbon. Absorbing agents often used in the petroleum industry, called molecular sieves, consist of porous aluminosilicates. Their porous structures have holes whose sizes are controlled to be the same as certain hydrocarbon molecules. Linear molecules are held in preference to branched ones. An example of this is in the separation of the C_4 refinery by-product mixture containing isobutene, 1-butene, 2-butene, and butane. *n*-Butenes and butane are absorbed in the 3–10 Å pores of the molecular sieve. The isobutene is cleanly separated, as it is too bulky to be absorbed. (See Weissermel and Arpe, 1978, p. 64.)

C. Crystallization

A crystalline substance is separated from a noncrystalline material by cooling a solution or a melt followed by filtering. With petroleum, waxes are removed in this way. Also, *p*-xylene can be separated from refinery streams containing the ortho and meta isomers by using this method.

D. Distillation

This process, which students carry out in first-year organic laboratory courses, gives us a way of separating molecules by size, because the number of carbon atoms in general is related to the boiling point. Passing petroleum once through a distillation column yields "straight-run fractions," as shown in Table 4.1. Recent work holds out the possibility of cracking crude oil without distillation (Wing, 1980).

4.1.2 Conversion Processes

Most refinery output is used for various kinds of gasoline. Only about 10% is used for petrochemicals. Straight-run gasoline, which contains C_4–C_{10} hydro-

TABLE 4.1 Data on Fractions from a Straight-Run Petroleum Distillation[a]

Name of Fraction	B.P. °C	Chemical Description
Gases	<20	C_1–C_4 paraffins, such as natural gas
Light naphtha (straight-run gasoline)	20–150	C_4–C_{10} aliphatic and cycloaliphatic compounds
Heavy naphtha	150–200	Used for fuel, chemicals
Kerosene	175–275	C_9–C_{16} compounds for jet and heating fuel
Gas oil	200–400	C_{15}–C_{25} compounds, for diesel and heating fuel and cracking to olefins
Lubricating oil	>350	Used for lubrication and catalytic cracking to lighter fractions
Heavy fuel oil	>350	Used for boiler fuel and catalytic cracking for lighter fractions
Asphalt		Used for paving, coating and structural purposes

[a] From H. A. Wittcoff and B. G. Reuben, *Industrial Organic Chemicals in Perspective,* Part 1, Wiley-Interscience, New York, 1980, p. 42.

carbons, has too low an octane number for high-compression engines. The octane number refers to the tendency of some hydrocarbons to ignite spontaneously in an engine. This results in a loss of power, because the gasoline burns on the upstroke compression phase. The straight-chain hydrocarbons show the greatest tendency, and the branched-chain hydrocarbons show the least. The octane number is assigned by comparing the sample of gasoline to mixtures of *n*-heptane and isooctane in which the *n*-heptane = 0 and isooctane = 100.

The aims of petroleum refining are as follows:

1. To produce as many high octane compounds in the gasoline as possible.
2. To change hydrocarbons whose size is greater than C_{10} or smaller than C_4 into substances in the gasoline range (C_4–C_{10}).

4.1.3 Types of Refinery Reactions

The processes in the following list have been developed in the last 50 years to separate crude petroleum, which contains hundreds of compounds, to yield a relatively small number of mixtures and compounds, which have well-defined uses. The procedures are as follows:

1. Thermal cracking
2. Catalytic cracking
3. Hydrocracking
4. Polymerization
5. Alkylation
6. Catalytic reforming
7. Interconversion of the seven basic chemicals

A. Thermal Cracking

Thermal cracking involves the use of heat (400–500°C) to cause conversion of saturates, that is, alkanes, into alkenes and hydrogen. Bonds from carbon to hydrogen are broken. The C–H bonds in paraffins are the most easily broken. The newest processes use 850–900°C and add steam to reduce the partial pressure of reactants and remove coke by making syngas.

Mechanism of a Thermal Cracking Reaction. We now discuss an example of how a cracking reaction proceeds. We use *n*-octane as an example. At cracking temperature, a C–C bond breaks and two free radicals form.

$$CH_3(CH_2)_6CH_3 \rightarrow CH_3CH_2CH_2CH_2CH_2\cdot + \cdot CH_2CH_2CH_3$$

An *n*-octane molecule loses a hydrogen to each radical, yielding a new free radical and an alkane of shorter chain length as shown in the following example:

$$CH_3CH_2CH_2CH_2CH_2\cdot + CH_3(CH_2)_6CH_3 \rightarrow$$
$$CH_3(CH_2)_3CH_3 + CH_3(CH_2)_5{-}\dot{C}H{-}CH_3$$

Note that the H abstraction occurs at a secondary carbon. All three radicals can undergo β-cleavage to give ethylene or propylene and a shorter radical.

$$CH_3CH_2CH_2CH_2CH_2\cdot \rightarrow CH_3CH_2CH_2\cdot + H_2C{=}CH_2$$

$$CH_3(CH_2)_3CH_2CH_2\dot{C}HCH_3 \rightarrow CH_3(CH_2)_3CH_2\cdot + H_2C{=}CHCH_3$$

(see Weissermel and Arpe, 1978, pp. 57–58).

B. Catalytic Cracking

Heating in the presence of catalysts not only accelerates the reaction rate of carbon to hydrogen bond breaking, but also causes carbon–carbon bond fracture.

This process is not a good source of petrochemicals, but is important for gasoline. The catalysts are called zeolites, molecular sieves on an aluminosilicate matrix, and are acidic catalysts. Reaction products different from those produced by thermal cracking are formed, such as aromatics and branched-chain molecules. The temperature is 450–550°C. An example of this is the cracking of gas oil (C_{15}–C_{25}) to "amylenes," a mixture of 2-methyl-1-butene and 2-methyl-2-butene. There are carbonium ion intermediates in this case.

C. Hydrocracking and Hydrotreating

In the presence of such catalysts as palladium on a zeolite and of hydrogen from dehydrogenation reactions, hydrocracking of paraffins yields two lower paraffins, and also olefins which do not undergo dehydrocyclization. These compounds are hydrogenated with or without isomerization. Using a Co–Mo catalyst on alumina, hydrocarbons with sulfur, nitrogen, and oxygen compounds yield ammonia, hydrogen sulfide, and water. The latter process is called *hydrotreating*.

D. Polymerization

Two low-molecular-weight olefins combine to yield a substance whose molecular weight is in the gasoline range. The catalyst is H_3PO_4 on clay. This process is not often used today for gasoline production, but is used with isobutene and also propylene to give C_8 or C_{12} olefin mixtures for detergent manufacture (see Figure 10.8).

E. Alkylation

In the case of alkylation, reaction of a low-molecular-weight olefin, for example isobutene, and of a paraffin such as isobutane can be used to yield hydrocarbon substances in the gasoline range. Friedel–Crafts catalysts such as sulfuric acid or HF are used (Wittcoff, 1980, p. 53). The mechanism of this reaction is given by Morrison and Boyd (1983, p. 372) and has carbonium ion intermediates. Because of the possibilities for various hydride abstractions and carbonium ion rearrangements, at least three major products are formed—2,2,4-trimethylpentane and the 2,3,4- and 2,3,3-trimethyl isomers. The branched products mean that the products have high octane numbers, so this process has major importance in refining.

F. Catalytic Reforming

Using Pt or Pt–Re on alumina we see dehydrogenation of both straight-chain and cyclic aliphatics to aromatics (benzene, toluene, xylene—BTX). Catalytic reforming is the most widely used refinery reaction. The carbon content of the molecules is generally not changed.

G. *Interconversion of the Seven Basic Chemicals*

The following reaction is an example where a less valuable member of this group (propylene) can be converted into two others in the series that are commercially more valuable (ethylene and butene).

$$2CH_2{=}CHCH_3 \rightarrow CH_2{=}CH_2 + CH_3CH{=}CHCH_3$$

This is a general reaction for many olefins and is called *metathesis* (see Section 4.6).

4.2 INDUSTRIAL PROCESSES USING ETHYLENE CHEMISTRY

We examine a few of the procedures used for processing organic chemicals produced in very large volume. These selected procedures show an approach that can be used for a great many reactions. Our discussion of each process will cover the following:

1. The most up-to-date process
2. Equipment needed
3. Mechanism of reaction
4. Why that particular process was developed

4.2.1 Replacement of One or More of the Hydrogens of Ethylene—Oxidation

We have chosen to discuss processes for the conversion of ethylene to vinyl chloride, acetaldehyde, and vinyl acetate. These processes are examples of catalytic reactions that recycle the catalyst and recycle the part of the reagent not incorporated into the product.

4.2.2 Vinyl Chloride (VCM)

Vinyl chloride is the monomer for the plastic PVC [poly(vinyl chloride)]. Production in the United States in 1984 was 7.51 billion pounds. This substance was formerly manufactured from acetylene. For acetylene manufacture, see section 4.8.3E

$$HC{\equiv}CH + HCl \xrightarrow[\text{on charcoal}]{HgCl_2} H_2C{=}CHCl$$

This process is very efficient, but the acetylene is so energy intensive that it was dropped. A more recent process is given in the following equation:

$$CH_2{=}CH_2 \xrightarrow[Cl_2]{FeCl_3} ClCH_2CH_2Cl \longrightarrow CH_2{=}CHCl + HCl \qquad (1)$$

A problem with this process is the generation of hydrogen chloride, which is nonreusable in this process. To overcome this, another process, called oxychlorination, has been developed. Ethylene is reacted with oxygen and hydrogen chloride using copper(II) chloride as catalyst. The equation is as follows:

$$CH_2{=}CH_2 + 2HCl + 0.5O_2 \rightarrow ClCH_2CH_2Cl + H_2O$$

The hydrogen chloride from the dehydrochlorination is available for reuse in the first step (1). This process was developed from a study of the Deacon reaction, which generates the chlorine as follows:

$$2HCl + 0.5O_2 \rightarrow Cl_2 + H_2O$$

The oxychlorination and the Deacon reaction use the same catalyst—copper(II) chloride. Process development work has shown that adding an equivalent amount of alkali metal chloride to reduce the volatility of the $CuCl_2$ and a small amount of rare earth chloride to overcome the rate decrease due to alkali greatly improves this process. These ingredients are mixed with an inert carrier and exist absorbed as a liquid at reaction temperature on the inert substance. The process is either a fixed or a fluidized bed procedure (see Chapter 11). Two recent reviews on this topic (Rylander, 1983, pp. 31–32, and Naworski and Velez, 1983) cover this process.

A. Mechanism of Vinyl Chloride Manufacture

The mechanism has recently been reviewed (Villasden and Livbjerg, 1978) and consists of three parts:

1. Oxychlorination:

$$CH_2{=}CH_2 + 2CuCl_2 \rightarrow ClCH_2CH_2Cl + 2CuCl \quad \text{(slow)}$$

2. Regeneration of catalyst:

$$CuCl + 0.5O_2 \rightarrow 0.5(ClCu)_2O_2 \quad \text{(fast)}$$

$$0.5(ClCu)_2O_2 + CuCl \rightarrow (ClCu)_2O$$

$$(ClCu)_2O + 2HCl \rightarrow 2CuCl_2 + H_2O$$

3. Direct chlorination. [See equation (1).]

$CuCl_2$ has been shown to chlorinate ethylene, and the temperature range of the reaction for the oxychlorination is 200–307°C. Considerable conversion occurs at 230°C. The Deacon process gives only an insignificant yield at this temperature, its reaction temperature being 300–400°C. It is suggested by Villasden and Livbjerg (1978) that absorption of ethylene on the copper(II) chloride is the rate-determining step for the oxychlorination at the favored temperature range.

B. Manufacturing Details for Vinyl Chloride

Figure 4.1 (Naworski and Velez, 1983, p. 242) shows a balanced vinyl chloride manufacturing flow sheet. It includes three stages, with two methods of chlorination—the direct method and the oxychlorination—both being used. It is a continuous process (see Section 8.3):

Direct Chlorination of Ethylene

$$CH_2{=}CH_2 + Cl_2 \rightarrow ClCH_2CH_2Cl$$

The temperature of chlorination is usually between 83 and 130°C. The catalyst consists of a low concentration of ferric chloride. Here the heat of chlorination is sufficient to purify the ethylene dichloride (EDC) by distillation.

Oxychlorination of Ethylene Figure 4.1 shows how the balancing of the two chlorination processes is achieved.

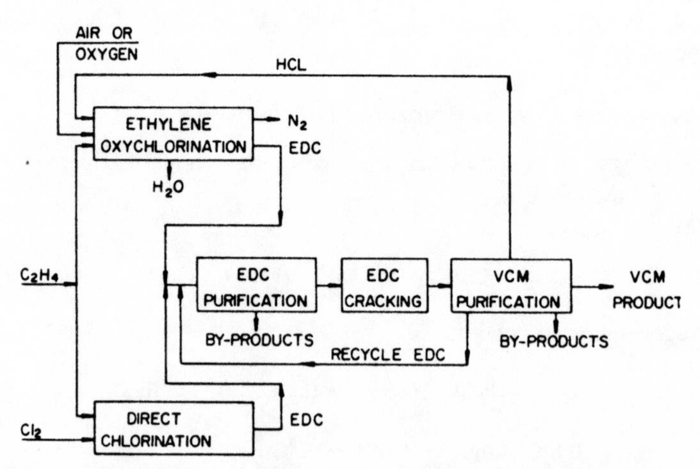

FIGURE 4.1 Balanced flow sheet for vinyl chloride manufacturing. (Reprinted with permission from J. S. Naworski and E. S. Velez, *Applied Industrial Catalysis*, B. Leach, Ed. Copyright, Academic Press, New York, 1983, Vol. 1, p. 242.)

$$CH_2{=}CH_2 + 0.5O_2 + 2HCl \rightarrow ClCH_2CH_2Cl + H_2O$$

The temperature for the oxychlorination reaction is 200–300°C and the pressure is 2–4 bar. It is an extremely exothermic reaction and requires efficient temperature control. Again, the heat of chlorination is sufficient to purify the EDC by distillation. In looking at Figure 4.2, we note a three-stage oxychlorination (vessels R_1, R_2, and R_3, filled with catalyst and connected in series) followed by an ethylene recovery unit R_4 (Naworski and Velez, 1983, p. 252). Here chlorine is added to convert the ethylene to EDC. The units are tubular heat exchangers. Each tube is filled with catalyst ($CuCl_2$ on an alumina support). The amount of catalyst on the support is low in the first reactor and increases in the later vessels. Most of the reaction and consequently most of the heat evolved occur in the first reactor. The EDC yield is 96% based on ethylene and 98% based on HCl.

Dehydrochlorination of Vinyl Chloride

$$ClCH_2CH_2Cl \xrightarrow{500\text{–}600°C} CH_2{=}CHCl + HCl$$

This reaction is conducted in a pyrolysis furnace and requires a large amount of heat. Tubular steel reactors are used with no catalyst. The EDC is in contact with the reactor for 5 seconds. About 98–99% of the converted EDC goes to vinyl chloride (VCM). Conversions of 50–60% are noted, and the unreacted EDC is recycled to the EDC purification vessel. The HCl is recycled to the oxychlorination reaction. Work is in progress on a lower cracking temperature

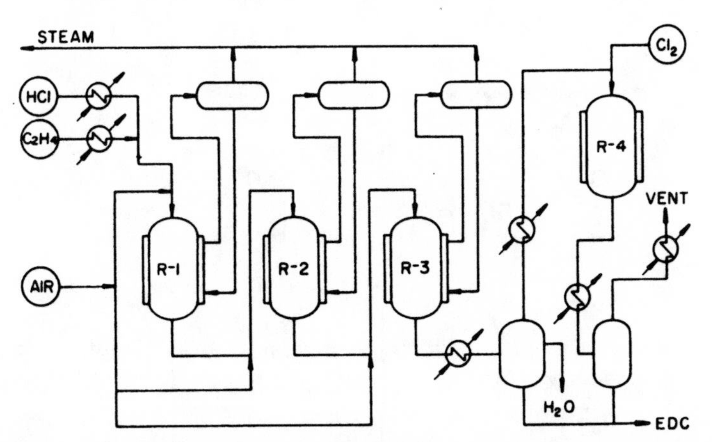

FIGURE 4.2 Detailed flow sheet for oxychlorination process for vinyl chloride. (Reprinted with permission from J. S. Naworski and E. S. Velez, *Applied Industrial Catalysis*, B. Leach, Ed., Copyright Academic Press, New York, 1983, Vol. 1, p. 252.)

with various catalysts. If this is successful, the process could then be conducted at the same throughput temperature as the oxychlorination.

4.2.3 Acetaldehyde

A recent review of acetaldehyde (Wittcoff, 1983) discusses the latest syntheses and uses of this product. We now examine the Wacker reaction, a cyclic process for manufacturing acetaldehyde. This process uses an equimolar amount of copper(II) chloride and a small quantity of palladium chloride as catalyst.

$$PdCl_2 + CH_2{=}CH_2 + H_2O \rightarrow CH_3CHO + Pd^0 + 2HCl \qquad (1)$$

$$Pd^0 + 2CuCl_2 \rightarrow PdCl_2 + 2CuCl \qquad (2)$$

$$2CuCl + 2HCl + 0.5O_2 \rightarrow 2CuCl_2 + H_2O \qquad (3)$$

Only a catalytic quantity of Pd is needed in this process because the $CuCl_2$ reoxidizes the Pd^0 as soon as it is formed [Eq. (2)]. The CuCl formed in the oxidation of the Pd is reoxidized to $CuCl_2$ by O_2 and HCl [Eq. (3)].

A. Mechanism of the Wacker Reaction

$$[PdCl_4]^{2-} + C_2H_4 \rightleftharpoons [PdCl_3(H_2C{=}CH_2)]^{-} + Cl^-$$

$$[PdCl_3(H_2C{=}CH_2)]^{-} + H_2O \rightleftharpoons PdCl_2(H_2C{=}CH_2)(OH_2) + Cl^-$$

$$PdCl_2(H_2C{=}CH_2)(OH_2) \rightleftharpoons [PdCl_2(H_2C{=}CH_2)(OH)]^{-} + H^+$$

The rate-determining step is the transformation of the hydroxy ethylene palladium complex into the β-hydroxy–ethyl palladium complex

$$[(C_2H_4)PdCl_2(OH)]^- + H_2O \rightarrow [(HO{-}CH_2{-}CH_2)PdCl_2(H_2O)]^-$$

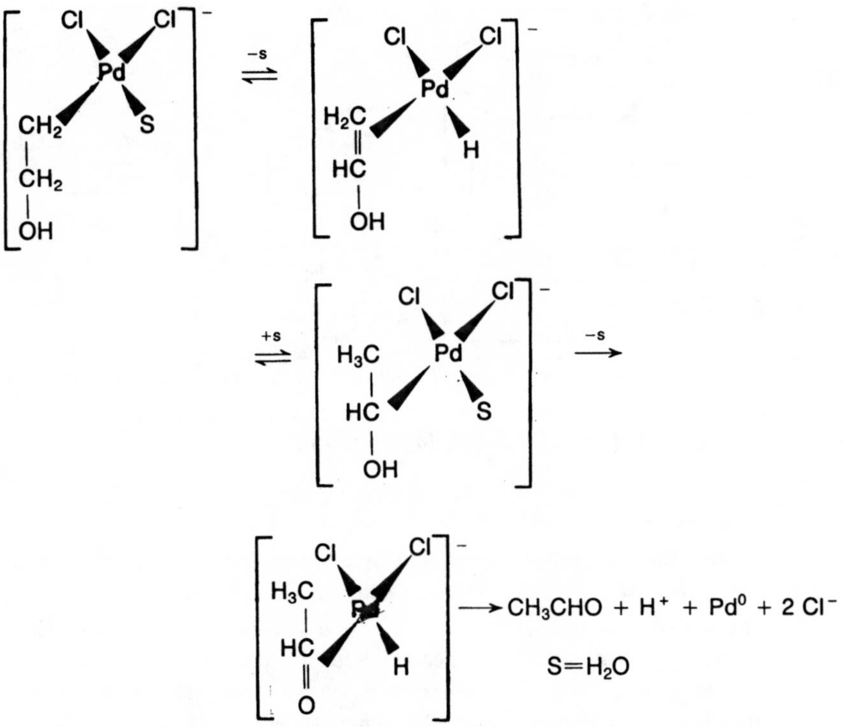

$$\longrightarrow CH_3CHO + H^+ + Pd^0 + 2\ Cl^-$$

$$S{=}H_2O$$

FIGURE 4.3 Mechanism of the Wacker reaction. (Reproduced with permission from R. Prins, *Chemistry and Chemical Engineering of Catalytic Processes*, R. Prins and G. Schuit, Eds., Copyright 1980 Martinus Nijhoff/Dr. W. Junk, Dordrecht, The Netherlands, p. 423, 425.)

Figure 4.3 shows the mechanism of the Wacker reaction. It will be discussed further in Section 11.2. This mechanism has been discussed by Prins (1980, pp. 423–425).

B. Manufacturing Conditions for Acetaldehyde (Wacker Method)

The process conditions for manufacturing acetaldehyde have been taken from the work of Jira et al. (1976, p. 97). See Figure 4.4.

1. Streams of ethylene and oxygen are separately piped into the bottom of a tower (A) containing copper(II) chloride solution mixed with a small amount of palladium chloride. The reactor is held at 130°C and 3 atm pressure.

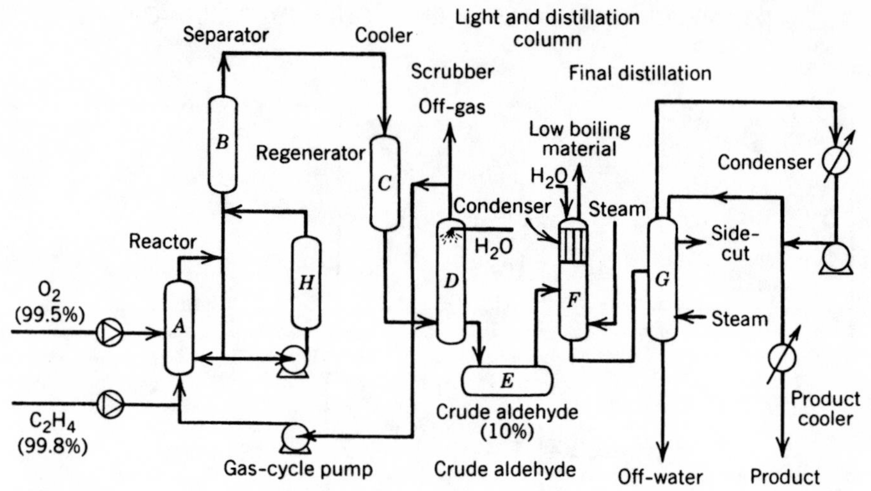

FIGURE 4.4 Acetaldehyde manufacturing flow sheet. (Reprinted with permission from *Hydrocarbon Processing*, Vol. 55, no. 3, March, 1976.)

2. From the reactor (A) is evolved acetaldehyde, water vapor, and unreacted ethylene and oxygen. These gases are sent to the demister-separator (B). Any catalyst solution carried as a mist is returned to (A).
3. The stream from (B) is cooled in (C) and sent to a scrubber (D). Here a water spray dissolves the acetaldehyde forming a solution which is carried to vessel (E). The remaining gases are recycled as shown to (A), after any consumed ethylene has been replaced.
4. Small amounts of reaction gas are removed to prevent buildup of contaminants.
5. Some of the catalyst solution is sent to the regenerator (H), along with sufficient oxygen to convert copper(I) to copper(II) chloride and decompose by-products. The temperature in (H) is 170°C.
6. The crude aqueous acetaldehyde is separated from low boiling material in a distillation column (F), which is operated as an extractive distillation with water. (See Section 9.1.1.)
7. Acetaldehyde is separated from water in a fractionating distillation column (G). High boiling material is removed as a higher boiling fraction or as distillation residue.

4.2.4 Reactions of Acetaldehyde

We now review procedures for the manufacture of various products that are carried out on an industrial scale using acetaldehyde as a raw material.

A. *Butanol*

Butanol has been prepared by three processes. The procedures were developed because of the need for cost reduction. Only reaction 2 uses acetaldehyde.

1. Butanol was first prepared by fermentation (see Section 3.1.5), a process that may become significant again, depending on petroleum costs.

2. When petroleum products became inexpensive, the process based on acetaldehyde became popular and was used for many years. This is given in the following equation:

$$2CH_3CHO \rightarrow CH_3CHOHCH_2CHO \rightarrow CH_3CH=CHCHO \xrightarrow{H_2} CH_3CH_2CH_2CH_2OH$$

3. The latest process, called hydroformylation, is discussed in Section 4.3.4. To prepare butanol, propylene is the raw material (Wittcoff, 1980).

$$CH_3CH=CH_2 + H_2 + CO \rightarrow CH_3CH_2CH_2CHO$$

$$\rightarrow CH_3CH_2CH_2CH_2OH + \text{isomer}$$

Propionaldehyde is the only product of hydroformylation of ethylene. This is a major raw material for propionic acid.

$$CH_2=CH_2 + H_2 + CO \rightarrow CH_3CH_2CHO \rightarrow CH_3CH_2COOH$$

B. *Acetic Acid and Peracetic Acid*

This process has been reviewed by Weissermel and Arpe (1978, pp. 152–153). Oxidation of acetaldehyde can be made to yield either acetic acid or peracetic acid, depending on the conditions.

$$CH_3CHO \rightarrow CH_3CO\cdot \xrightarrow[O_2]{} CH_3COOO\cdot \rightarrow \underset{\text{peracetic acid}}{CH_3COOOH} \quad (1)$$

$$CH_3CHO + CH_3COOOH \rightarrow CH_3CHOHOOCOCH_3 \rightarrow 2CH_3COOH \quad (2)$$

With mild conditions (15–40°C) at 25 bar and air, one gets peracetic acid as the main product. To make acetic acid, the procedure is to accelerate the oxidation by radical formation as well as decomposition of peracetic acid. The catalysts used are Mn or Co acetates. The catalyst enters into the reaction as shown:

Initiation of reaction

$$CH_3CHO + Mn^{3+} \rightarrow CH_3\underset{\underset{O}{\parallel}}{C}\cdot + Mn^{2+} + H^+$$

Acceleration of peracetic acid decomposition

$$CH_3\underset{\underset{O}{\parallel}}{C}{-}OOH + Mn^{2+} \rightarrow CH_3\underset{\underset{O}{\parallel}}{C}O\cdot + Mn^{3+} + OH^-$$

$$CH_3\underset{\underset{O}{\parallel}}{C}{-}O\cdot + CH_3CHO \rightarrow CH_3\underset{\underset{O}{\parallel}}{C}{-}OH + CH_3\underset{\underset{O}{\parallel}}{C}\cdot$$

A new acetic acid process (developed by Monsanto), not involving acetaldehyde, involves the carbonylation of methanol to yield acetic acid:

$$CH_3OH + CO \rightarrow CH_3COOH$$

As is shown later, this procedure is related to the hydroformylation process. Rhodium is preferred to cobalt as catalyst (Wittcoff, 1983). The catalyst is rhodium chloride, iodine promoted. Cobalt acts as a ligand for the rhodium. See Figure 4.5 for the mechanism of this reaction. This process has been recently reviewed by Eby and Singleton (1983, pp. 275–296) and Forster and Dekleva (1986) and is discussed in Chapter 11.

4.2.5 Preparation of Methanol

Since methanol has been used in the acetic acid synthesis just described, this seems an appropriate point to discuss methanol production. It will be mentioned later on at several points. Methanol is made from syngas as shown in the following equation:

$$CO + 2H_2 \rightarrow CH_3OH$$

The production of methanol in the United States was about 8.28 billion pounds in 1984. This synthesis has recently been discussed by Marschner and Moeller (1983). It should be considered along with other reactions involving the hydrogenation of carbon monoxide. Among these are methanation (Section 2.3.2) and Fischer–Tropsch synthesis (Section 4.8.3.A) as well as higher alcohols, dimethyl ether, esters, ketones, and aldehydes. The above authors pointed out

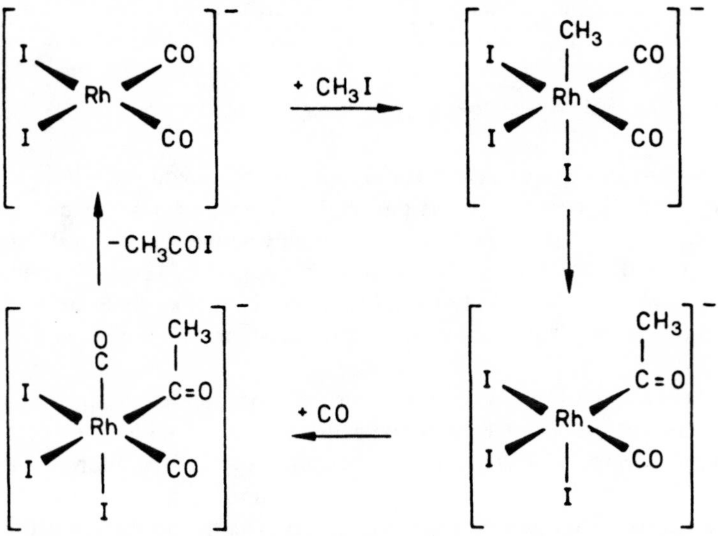

$+ CH_3I$

$-CH_3COI$

$+ CO$

$$CH_3COI + CH_3OH \longrightarrow CH_3COOH + CH_3I$$

FIGURE 4.5 Mechanism for the conversion of methanol to acetic acid. (Reprinted with permission from R. Eby and T. Singleton, *Applied Industrial Catalysis*, B. Leach, Ed., Copyright Academic Press, New York, 1983, Vol. 1, p. 282.)

that methanol is the least thermodynamically stable of these products from CO and hydrogen formed under similar conditions.

A. Methanol process notes

Marschner and Moeller (1983) point out that the formation of methanol is highly exothermic. Synthesis gas plants making methanol employ zinc oxide–chromium oxide catalysts. The process until the early 1970s operated at approximately 300 bars and at temperatures of 320–400°C. About this time relatively sulfur-free (0.5 ppm) synthesis gas became available. Zinc oxide or molecular sieves is used for this, preceded by hydrogenation if necessary. The new conditions are pressure 50–100 bar and temperature 230–300° and a change to highly active copper catalysts. The chemistry is explained in the following three equations (Marschner and Moeller, 1983, pp. 217–218):

$$CO + 2H_2 \rightarrow CH_3OH \tag{1}$$

$$CO_2 + 3H_2 \rightarrow CH_3OH + H_2O \tag{2}$$

$$CO + H_2O \rightarrow CO_2 + H_2 \tag{3}$$

The synthesis gas from methane has the composition shown by the following equation:

$$CH_4 + H_2O \rightarrow CO + 3H_2 \quad (4)$$

The synthesis gas has to have a certain amount of carbon dioxide in it to use up some of the hydrogen. This is prepared by reaction (3) or some is added to the process stream. The catalyst used is a highly active copper, freshly reduced from CuO with H_2. Methane formation is suppressed by keeping the quantities of Fe, Co, and Ni in the catalyst at a low value. Higher alcohols are suppressed by controlling the alkali and alkaline earth metals in the catalyst at essentially zero.

Figure 4.6 shows two process streams for producing methanol from natural gas. In the first process, CO_2 must be added to adjust the hydrogen content, as explained in the previous paragraph. The temperature of the reforming reaction is 780°C. In the second process, the gas is reformed in the presence of oxygen (see partial oxidation part of Section 2.3.1.A). The temperature is 950°C. The resulting syngas pressure is increased at this temperature, and so less energy is needed to reach the pressure needed for methanol synthesis.

4.2.6 Vinyl Acetate

The preparation of vinyl acetate is basically similar to that for acetaldehyde—a catalytic process, where the oxidizing agent is palladium ion. If the reaction is conducted in a solution of acetic acid with added sodium acetate, acetoxylation

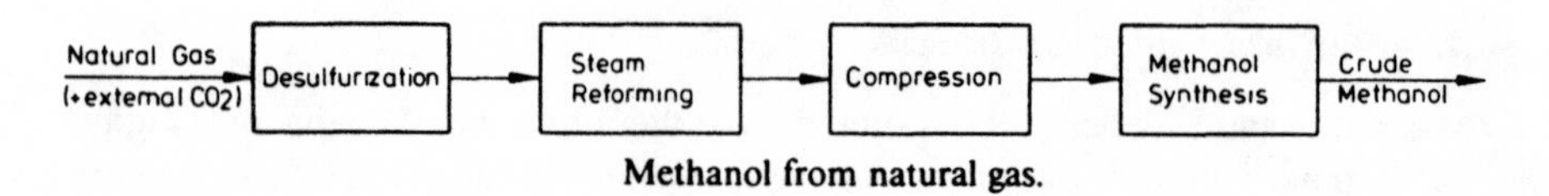

Methanol from natural gas.

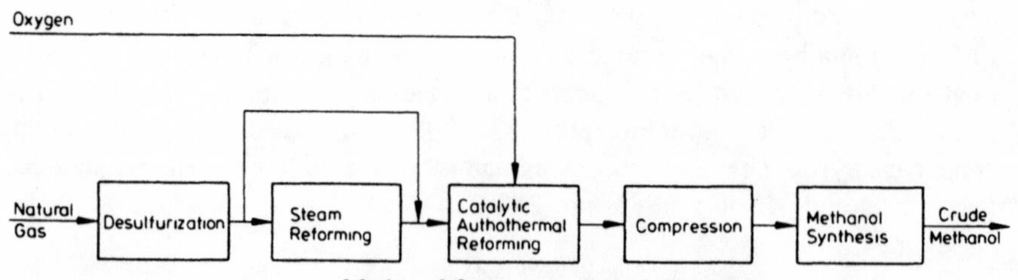

Methanol from natural gas and oxygen.

FIGURE 4.6 Methanol process flow diagrams from methane. (Reprinted with permission from F. Marschner and F. W. Moeller, *Applied Industrial Catalysis*, B. Leach, Ed. Copyright, Academic Press, New York, 1983, Vol. 2, p. 229.)

takes place. The regeneration of the Pd^0 to Pd^{2+} by copper(II) chloride and the recovery of the CuCl by oxygen and HCl are the same as with acetaldehyde (Weissermel and Arpe (1978, p. 203). The process is vapor phase (Gates et al., 1979) controlled so that the acetaldehyde, a by-product in this process, can be oxidized in the reactor to supply acetic acid raw material.

$$CH_2{=}CH_2 + PdCl_2 + 2CH_3COONa \rightarrow$$

$$CH_2{=}CHOCOCH_3 + 2NaCl + Pd^{\circ} + CH_3COOH$$

4.2.7 Reactions Where Addition to the Ethylene Double Bond Occurs

We now discuss processes for a few of the addition reactions of ethylene. Again all are catalytic procedures. The substances covered are as follows:

Ethylene oxide
Ethylene glycol
Styrene
Propionaldehyde
Ethanol

4.2.8 Ethylene Oxide

The U.S. production of ethylene oxide (EO) in 1984 was 5.96 billion pounds. The manufacture was carried out essentially by air or oxygen oxidation of ethylene over a silver catalyst.

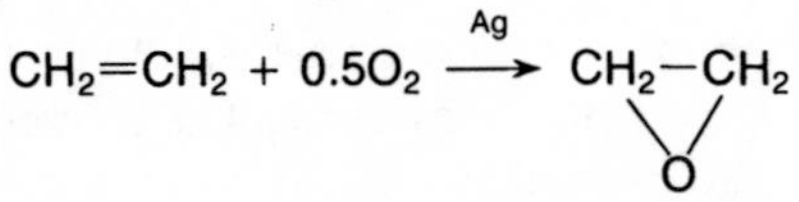

$$CH_2{=}CH_2 + 0.5O_2 \xrightarrow{Ag} \underset{\diagdown\ \diagup \atop O}{CH_2{-}CH_2}$$

The catalyst is 10–15% silver on a support of low-surface alumina. This support disperses the metal on its surface as small crystallites. The surface is large compared to the amount of metal used. The support dissipates heat and increases resistance to poisoning.

Silver has its electronic properties changed by contact with the support. Ethylene does not absorb well on reduced clean silver, but is taken up readily on silver containing preabsorbed molecular oxygen. It is considered that ethylene is attached to the layer of molecularly absorbed oxygen (O_2 ads), and in this form reacts to form ethylene oxide. The following equation is taken from Rylander (1983, p. 24). A review of this process is given by Berty (1983).

$$\begin{array}{ccccc} CH_2{=}CH_2 & & CH_2{-}CH_2 & & CH_2{-}CH_2 \\ & & \diagdown\ \diagup & & \diagdown\ \diagup \\ O^{\delta\oplus} & & O^{\oplus} & & O \\ \| & & | & & \\ O^{\delta\ominus} & \longrightarrow & O^{\ominus} & \longrightarrow & O \\ | & & | & & | \\ Ag & & Ag & & Ag \end{array}$$

The remaining absorbed oxygen atoms (O ads) react with ethylene to give CO_2 and H_2O.

$$6O_{(ads)} + CH_2{=}CH_2 \rightarrow 2CO_2 + 2H_2O$$

Using this approach, the overall reaction would be

$$7C_2H_4 + 6O_{2(ads)} \rightarrow 6C_2H_4O + 2CO_2 + 2H_2O$$

The maximum yield by this mechanism would be 6/7 or 85%, the theoretical maximum yield of ethylene oxide that can be reached by this mechanism. In practice this yield is seldom reached (Rylander, 1983, p. 24). The results given here show that it is valuable to understand the mechanism of a reaction. In this case the experimenter cannot expect a better yield unless the conditions at the catalyst surface are altered. It has been found that the addition of a few ppm of ethylene dichloride appears to alter the ratio of molecular and atomic oxygen at the catalyst surface, producing a consistently higher yield. When there are ensembles of four Ag atoms on the surface (see Section 11.1.2.B), the conversion to CO_2 is favored. The added chlorine is considered to reduce the number of these ensembles, and favor EO yield.

An older process, the reaction of ethylene with hypochlorous acid, was the preferred method of synthesis for many years. The chlorohydrin intermediate is decomposed with calcium hydroxide:

$$CH_2{=}CH_2 + HOCl \rightarrow CH_2OHCH_2Cl + Ca(OH)_2 \rightarrow \underset{\diagdown O \diagup}{CH_2{-}CH_2} + CaCl_2$$

This procedure has two major drawbacks:

1. There is a disposal problem for huge quantities of calcium chloride.
2. The chlorine added as HOCl is lost, not being used in the product.

4.2.9 Ethylene Glycol

The U.S. production of ethylene glycol (EG) in 1984 was 4.84 billion pounds. This product has been the major outlet for ethylene oxide.

$$\underset{\diagdown O \diagup}{CH_2CH_2} + H_2O \rightarrow HOCH_2CH_2OH$$

The process uses 10 molar excess of water with an acidic catalyst at 50–70°C, or catalyst-free at 140–230°C and 20–40 bar. The selectivity is only 90%, the rest being polyethylene glycols of various chain lengths. This is a low selectivity for a product made in such large quantities.

New synthetic paths for the glycol will make this intermediate much less important. These new routes are discussed in a recent review (Kollar, 1984). Five routes are under active investigation.

A. *Raw Materials for Ethylene Glycol*

The following are the chief raw materials used in the various syntheses:

1. Carbon monoxide and hydrogen. This mixture, called syngas, has to have a low sulfur content because of the use of noble metal catalysts. High pressures are usually needed. Some of the procedures require separation of the syngas into its components. For background see Sections 2.3.1.A and 4.2.5.A.

2. Methanol. The cost of producing methanol from syngas is very low, and so it compares as a raw material with this mixture. See Section 4.2.5.A.

$$CO + 2H_2 \rightarrow CH_3OH$$

3. Formaldehyde. There are two syntheses:

Oxidation

$$CH_3OH + 0.5O_2 \rightarrow HCHO + H_2O$$

The catalyst used is Fe_2O_3–MoO_3 or silver

Dehydrogenation

$$CH_3OH \rightarrow HCHO + H_2$$

The catalyst used is silver. The reaction is driven to the right by burning the hydrogen as fuel. This raw material has a wide range of costs, depending on purity. Any form of anhydrous HCHO is too expensive at present for EG.

B. Processes for Production of EG—All Based on Syngas

The following reactions are given as an example of how many different processes may be developed when there is an economic incentive to do so.

Direct Reaction. This procedure is at first glance the most straightforward, but really is not since practically every bond in the product molecule has to be formed. It consists of the rhodium-catalyzed condensation of syngas at 3400 atm as shown in the following equation:

$$2CO + 3H_2 \rightarrow \text{purified syngas} \rightarrow HOCH_2CH_2OH \rightarrow \text{purified EG}$$

This process is reviewed by Dombek (1986).
This process has the following problems:

1. Rhodium catalyst requires an extremely low sulfur syngas.
2. High pressure is needed.
3. A low rate of reaction is seen.

Carbonylation. In this process formaldehyde is reacted with more CO. This is followed by a reduction as shown. This is the oldest of the syngas EG syntheses.

$$HCHO \xrightarrow[HF,\ H_2O]{CO} \underset{\text{glycolic acid}}{HOCH_2COOH} \xrightarrow{CH_3OH} HOCH_2COOCH_3$$

$$\xrightarrow{H_2} HOCH_2CH_2OH$$

This process has the following problems:

1. The HCHO must be methanol free.
2. Syngas components have to be separated, adding to the cost.
3. Syngas is used as the CO source for glycolic acid formation. This separates the CO from the H_2, which is used for the reduction step.

Hydroformylation. The hydroformylation procedure is shown in Section 4.3.4. It is the reaction of formaldehyde with $CO + H_2$, followed by hydrogenation. This procedure, where HCHO is subjected to hydroformylation conditions, is discussed in a review by Dombek (1986). Normally the substrate is an olefin.

$$HCHO + H_2 + CO \rightarrow HOCH_2CHO + H_2 \rightarrow HOCH_2CH_2OH$$

This process has the following problems:

1. $RhCl(CO)[(C_6H_5)_3P]_2$ catalyst is expensive and needs low sulfur content in both HCHO and syngas.
2. The form of HCHO needed is the para form or the trioxane, which are expensive. With aqueous formaldehyde, the EG yield is quite low.
3. The reaction rate is slow unless the procedure is run at high pressure (4000 psi).

Reductive Hydroformylation. In this reaction, reduction and the hydroformylation are combined in one reactor. The catalyst used is $Rh(CO)_2$(acetylacetonate) in *N*-methyl pyrrolidone.

$$HCHO + H_2 + CO \rightarrow HOCH_2CH_2OH$$

This procedure has the same problems as the preceding process.

Oxidative Coupling As one can see from the following equation, some of the synthesis gas components have to be separated:

$$CO + O_2 + C_2H_5OH \xrightarrow[CuCl_2]{PdCl_2} C_2H_5OOCCOOC_2H_5 + H_2 \rightarrow HOCH_2CH_2OH$$

The first step uses a palladium catalyst. The Pd has to be reoxidized. Copper salts are used for this, as well as alkyl nitrite. This procedure has the following problems:

1. Syngas components must be separated, so raw material costs are high.
2. The CO has to be free of hydrogen which if present gives water, which is detrimental to oxidation.
3. Reduction of the diethyl oxalate takes extremely high pressure.

4.2.10 Styrene

The U.S. production for this monomer in 1984 was 7.71 billion pounds.

A. Classical Reaction for Styrene

Ethyl benzene is dehydrogenated to styrene as shown in the following reaction. Ethyl benzene is prepared from benzene and ethylene by Friedel–Crafts catalysts, such as H_3PO_4:

$$C_6H_6 \xrightarrow{CH_2=CH_2} C_6H_5CH_2CH_3 \xrightarrow[\text{superheated steam}]{FeO \cdot K_2CO_3} C_6H_5CH=CH_2 + H_2$$

Dehydrogenation is brought about at 550–600°C over an iron oxide–potassium carbonate catalyst in a tank-type reactor. Incoming steam not only supplies heat, maintains the catalyst iron in a highly oxidized state, and reduces the partial pressure of ethyl benzene to limit side reactions. The waste hydrogen is used as fuel gas for the generation of heat for the dehydrogenation.

The problems with this reaction are as follows:

1. The purification is difficult compared with the oxirane process described below.
2. The reaction temperature is much higher than in the oxirane procedure, therefore it is more difficult to avoid polymerization because of longer cooling time.

B. *The Oxirane Process–The Modern Method*

The most up-to-date process for producing styrene monomer is called the oxirane process. The approach to this synthesis is the use of a reagent that will produce another commercially valuable product along with styrene. Labor costs per kilogram of saleable product are clearly reduced. See Chapter 6. This synthesis device is used in several commercial processes, a few of which are discussed here.

In this process ethyl benzene is converted to its hydroperoxide. The latter is reacted with propylene, giving propylene oxide and methylphenyl carbinol, which is dehydrated to styrene (Weissermel and Arpe, 1978, p. 237). See Figure 4.9 and Section 4.3.2.A for the reaction scheme for styrene formation by the oxirane process (Landau et al., 1979, p. 602). This reaction results in two commercial products in one reactor—styrene and propylene oxide.

4.2.11 Propionaldehyde

Propionaldehyde is prepared from ethylene, hydrogen, and carbon monoxide. This is an example of a *hydroformylation* reaction, a general reaction of all olefins using a 1 : 1 mixture of hydrogen and carbon monoxide. The double bond can be terminal or internal. The H_2–CO mixture adds to the olefin as —H and —HCO and products from both modes of addition are seen.

The equation for propionaldehyde is as follows. Note that only one product is obtained with ethylene, but at least two with all other olefins.

$$CH_2{=}CH_2 + H_2 + CO \rightarrow \underset{\text{propionaldehyde}}{CH_3CH_2CHO}$$

With most higher olefins, the double bond can migrate under reaction conditions, and products from each olefin are found. The following equation shows how the reaction proceeds with higher olefins.

$$CH_3CH{=}CH_2 + H_2 + CO \rightarrow CH_3CH_2CH_2CHO + CH_3\underset{\displaystyle CHO}{\underset{|}{C}}HCH_3$$

The straight-chain product is present in largest yield and can be favored by proper choice of catalyst. The catalysts used are cobalt, rhodium, or ruthenium compounds, modified with amine or phosphorus ligands. The currently used mechanism is discussed in Section 4.3.4A. Hydroformylation is basically a homogeneously catalyzed reaction, taking place at 200–450 bar and 100–200°C.

4.2.12 Ethanol

This reaction is carried out in the gas phase in the presence of acidic catalysts such as H_3PO_4 on SiO_2. The conditions of 300°C, 70 bar, and a short reaction time have been found to reduce the formation of diethyl ether and ethylene oligomers. A water/ethylene mol ratio of 0.6–1 is necessary to prevent loss of catalytic activity and loss of H_3PO_4. The ethylene selectivity is 97% but the conversion is only 4%. Ethylene has to be recycled, and the alcohol is purified by extractive distillation (see Section 9.1.1) (Weissermel and Arpe, 1978, pp. 172–173). Fermentation is the old way to prepare ethanol and remains as the backup in the event of an ethylene shortage. See Section 3.1.1.A.

$$CH_2{=}CH_2 + H_2O \rightarrow CH_3CH_2OH$$

4.2.13 Polymerization of Ethylene

High-density polyethylene is obtained with metal alkyl or metal oxide catalysts. Hydrocarbon solutions of $TiCl_4$ are used in the presence of $Al(C_2H_5)_3$ to make this high-density material. The mechanism has recently been reviewed by Henrici-Olive and Olive (1981) and is shown in Figure 4.7. A shows the catalyst mixture, to which is coordinated one of the ethylene molecules about to be polymerized. The chain growth occurs by mechanism a or b via the intermediate B to give products C_1 or C_2. The latter structures show an additional molecule of ethylene coordinated, ready to be added to the chain.

This is the only polymer process mechanism to be mentioned in this book. As previously stated, the author recommends that an additional one-semester course on polymers be taken concurrently or later.

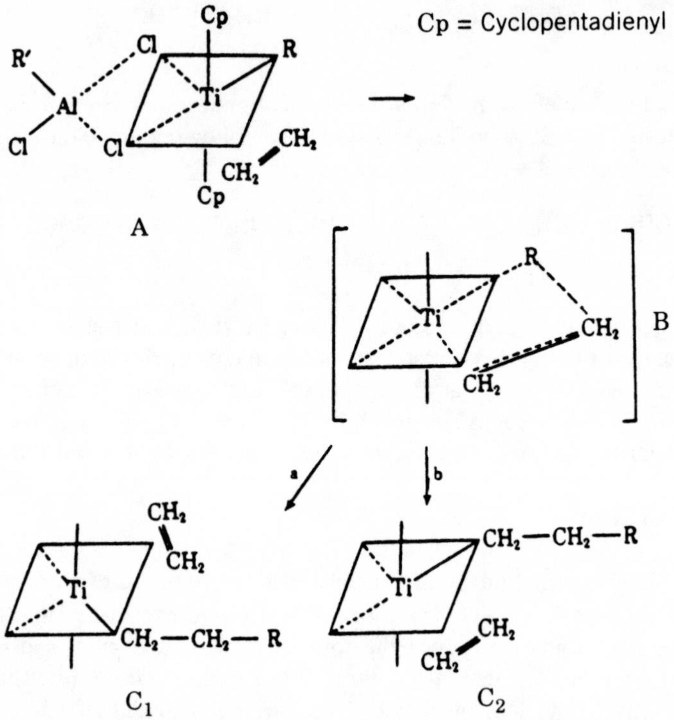

Cp = Cyclopentadienyl

FIGURE 4.7 Mechanism for the polymerization of ethylene. (Reprinted with permission from G. Henrici-Olive and S. Olive, *CHEMTECH*, **11**, 749. Copyright 1981 American Chemical Society.)

4.3 INDUSTRIAL PROCESSES USING PROPYLENE CHEMISTRY

We have studied some of the reactions of interest to industry with ethylene as a raw material, where one hydrogen is replaced by chlorine, acetoxy, or hydroxy. As we consider propylene, the replacement reactions seen are those where the three allylic hydrogens are reactive. For available raw material for propylene, we have two sources:

1. Thermal cracking of natural gas, which contains propane and ethane, yields propylene and ethylene.
2. Catalytic cracking of higher petroleum fractions yields propylene with other products.

4.3.1 Oxidation Reactions of Propylene

We first discuss the preparation of two substances, acrolein and acrylonitrile, prepared from propylene by similar catalytic processes.

A. Background on Reactions for Manufacturing of Acrolein and Acrylonitrile

Acrolein is used as a raw material for acrylic acid and other large-volume organics. The United States had a production in 1984 of 2.20 billion pounds. Essentially all of this is prepared by the following process.

The equation for acrylonitrile manufacture is:

$$CH_3CH{=}CH_2 + NH_3 + 1.5O_2 \xrightarrow[400\text{–}460°C]{} CH_2{=}CHCN + 3H_2O$$

and for acrolein production:

$$CH_3CH{=}CH_2 + O_2 \xrightarrow[300\text{–}450°C]{} CH_2{=}CHCHO + H_2O$$

Both of these processes use a catalyst containing bismuth and molybdenum oxides along with a variety of other components. It should be noted that five other processes for acrylic acid have been used previously (Wittcoff, 1980, p. 72). The new catalytic processes were developed because of the ever-present economic need for lower cost. The processes using the $Bi_2O_3 \cdot nMoO_3$ are discussed in a recent paper by Grasselli and Burrington (1984, p. 393–404), and by Grasselli (1986, pp. 216–221). In Figure 4.8*a*, Grasselli and Burrington show the three possible structures for the common intermediate and in their paper explain why the allylic was chosen. Figure 4.8*b* shows the simplified process showing how O or N is added to the common intermediate to give the two products. Figure 4.8*c* shows the detailed structure of the bismuth–molybdenum complex as it interacts with its propylene to give the two products. This detailed work is justified by the commercial success of the process. There are two stages:

1. Slow reduction of the catalyst metal oxide:

$$2M^{n+}O_x + CH_2{=}CHCH_3 \rightarrow 2M^{(n-4)}O_{x-2} + CH_2{=}CHCHO + H_2O$$

2. Fast replacement of catalyst lattice oxygen:

$$2M^{n-4}O_{x-2} + O_2 \rightarrow 2M^{n+}O_x$$

A large excess of air is used for acrolein manufacture and a slight excess of air for the acrylonitrile process.

B. Manufacturing of Acrylonitrile

The reaction of an olefin with ammonia and oxygen is called ammoxidation. Figure 4.8*d* shows one of several equipment setup modifications for acrylonitrile

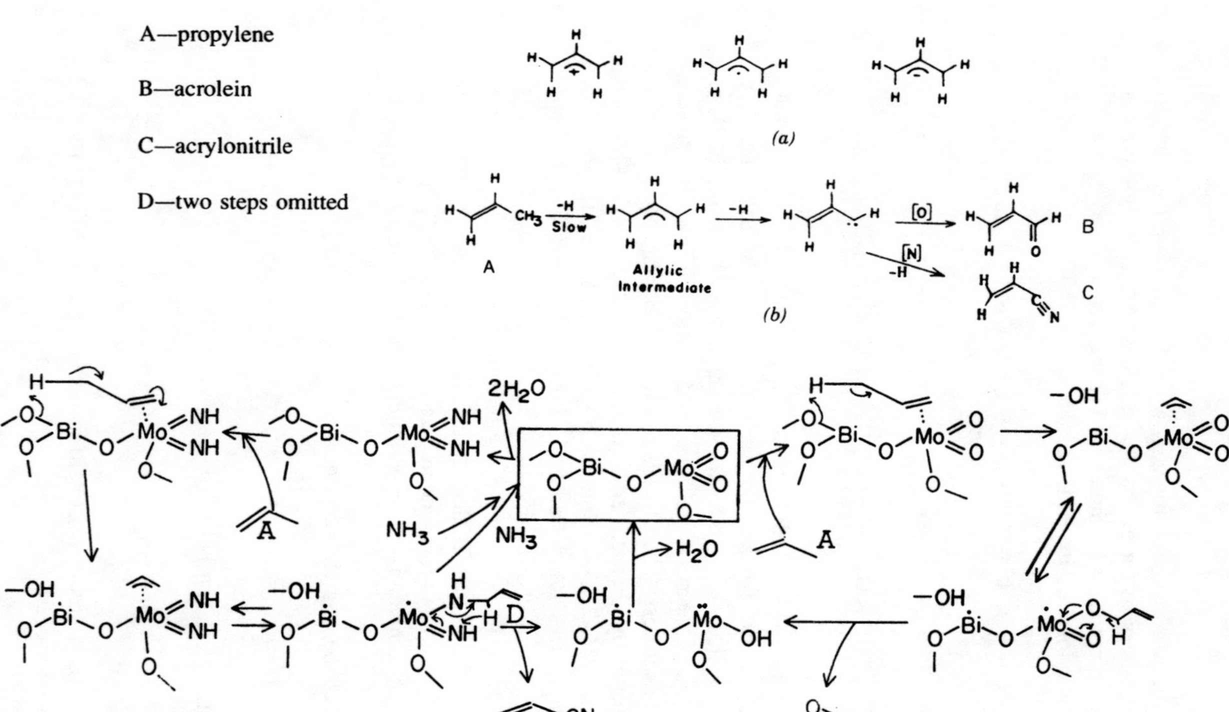

FIGURE 4.8*a–c*: Preparation of acrylonitrile and acrolein. (*a*) structures possible for allylic intermediate; (*b*) equation for the preparation; (*c*) detailed mechanism for the preparation, showing the reaction of the $Bi_2O_3 \cdot nMoO_3$ catalyst with propylene. (Reprinted with permission from R. D. Grasselli and J. D. Burrington, *Industrial and Engineering Chemistry—Product Research and Development*, **23**, 393, 402 (1984). Copyright American Chemical Society.)

The acrylonitrile process

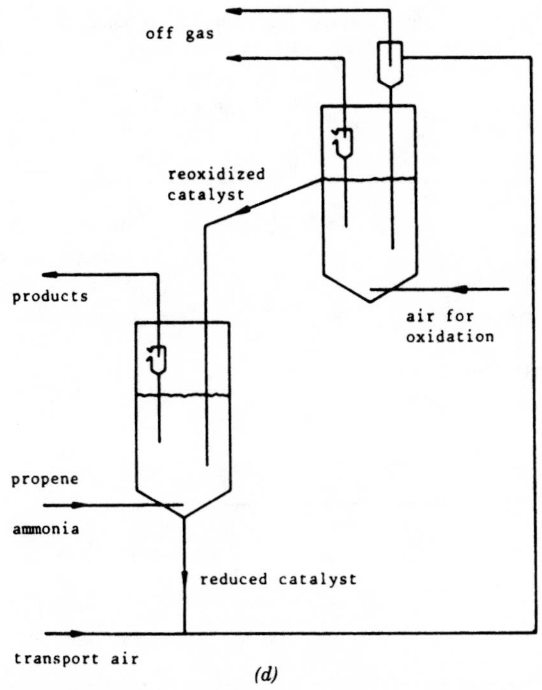

FIGURE 4.8*d* Acrylonitrile manufacturing flow diagram. (Reproduced with permission from H. S. van den Baan, *Chemistry and Chemical Engineering of Catalytic Processes*, R. Prins and G. Schuit, Eds. Copyright 1980, Martinus Nijhoff/Dr. W. Junk, Dordrecht, The Netherlands, p. 527.)

manufacture. There are two vessels one for each process step. (See equations 1 and 2 above). Other equipment schemes are given in van der Baan (1980, pp. 523–527). The equipment uses fluidized bed catalysis (see page 208). Equipment is now available to do both steps in one reactor.

The propylene molecule must lose three hydrogens and the dehydrogenated intermediate must be bonded to the catalyst in such a way that these intermediates are protected against direct oxidation by gaseous oxygen (van Hooff, 1980, pp. 512–513; Stiles, 1983). The design of the equipment shows one equipment design which illustrates how the 2 process stages are carried out.

The by-products are H_2O, CO–CO_2, CH_3CHO, N_2, CH_3CN, and HCN. The latter two have commercial value and are generally recovered from the product stream (see below). The catalyst is stable and can be used without much loss of activity if the temperature is kept below 525°C. For the recovery method for

TABLE 4.2 Work-up Process for Acrylonitrile

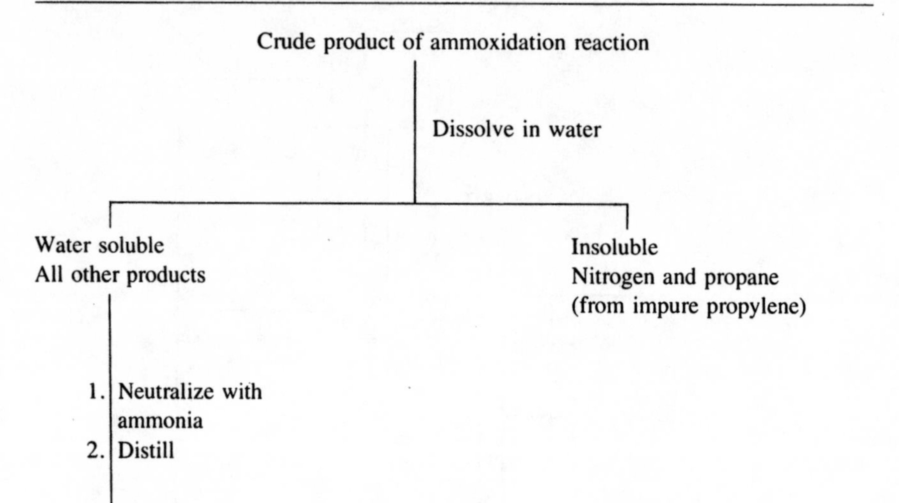

Acrolein + Resin

Distillate

Acrylonitrile, 1000 kg
Acetonitrile, 30–40 kg
Hydrogen cyanide, 140–180 kg

acetonitrile and hydrogen cyanide, see Table 4.2. The hydrogen cyanide, although a by-product, is recovered in such large quantities that it becomes a major source of this basic raw material. See Section 4.8.3.B. The catalyst studies are continuing for more selectivity.

4.3.2 Propylene Processes Giving Two Commercial Products

Propylene processes put on stream by Halcon International Corp. involve the preparation of propylene oxide (PO) along with one other commercial product in the same reactor. This procedure is called the oxirane process and is a homogeneous catalysis procedure.

A. Example 1: A Two-Product Process—Preparation of Styrene (S) and Propylene Oxide (PO)

Description of the oxirane process (see Figure 4.9).

1. Ethyl benzene is oxidized to the hydroperoxide by air or oxygen in the presence of an initiator. Molybdenum, vanadium, or technetium, as their

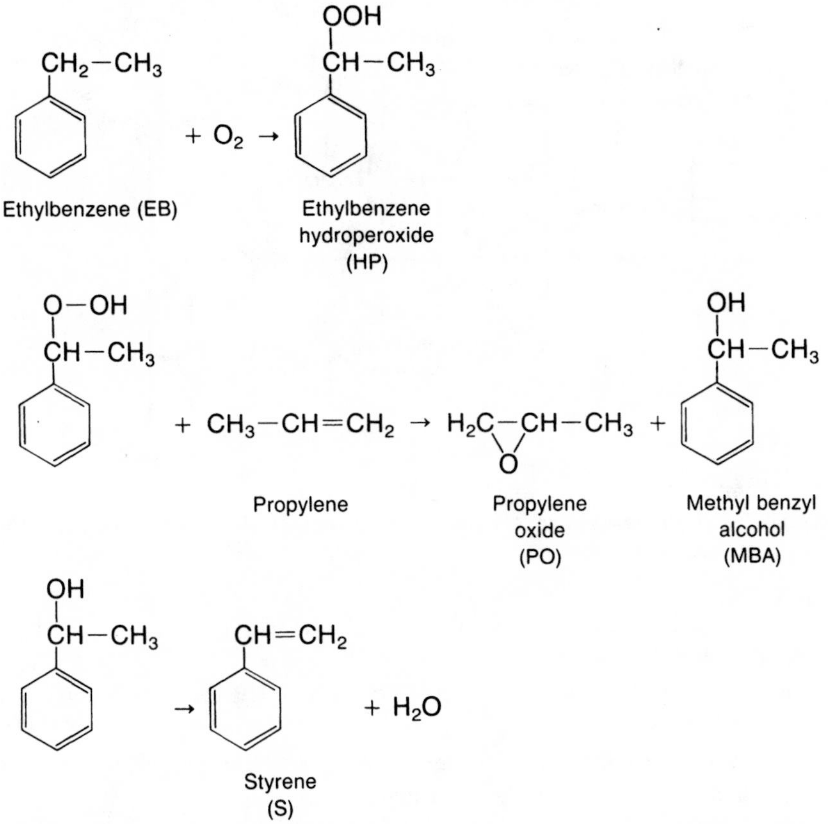

$C_6H_5CH_2{-}CH_3 + O_2 \rightarrow C_6H_5CH(OOH){-}CH_3$

Ethylbenzene (EB) | Ethylbenzene hydroperoxide (HP)

$C_6H_5CH(O{-}OH){-}CH_3 + CH_3{-}CH{=}CH_2 \rightarrow H_2C{-}CH{-}CH_3\ (\text{epoxide, }O) + C_6H_5CH(OH){-}CH_3$

Propylene | Propylene oxide (PO) | Methyl benzyl alcohol (MBA)

$C_6H_5CH(OH){-}CH_3 \rightarrow C_6H_5CH{=}CH_2 + H_2O$

Styrene (S)

FIGURE 4.9 Styrene and propylene oxide produced by the oxirane process: reaction scheme. [Reprinted with permission from R. Landau et al., *CHEMTECH* **9,** 604 (1979). Copyright American Chemical Society.]

naphthenates are used for this purpose. These materials are soluble in hydrocarbons.

2. The crude hydroperoxide epoxidizes propylene, and the propylene oxide is distilled off.
3. Methyl benzyl alcohol is converted into styrene at atmospheric pressure in the gas phase (180–200°C), using TiO_2 or Al_2O_3 as catalyst.

S and PO are two of the 50 largest-volume chemicals produced in the United States.

A flow chart used for this preparation is shown in Figure 4.10. *Manufacturing details* styrene and propylene oxide:

1. EB is oxidized to HP in a stream of air in reactor 1.

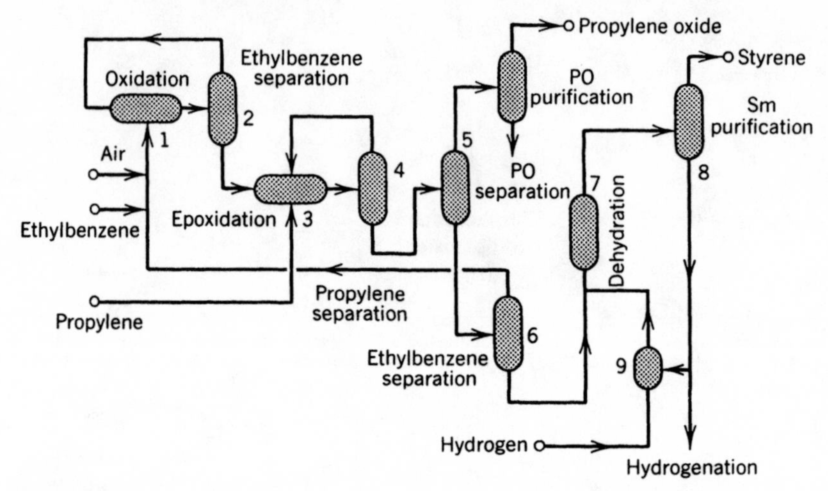

FIGURE 4.10 Styrene and propylene oxide by the oxirane process: flow diagram. (Reprinted with permission from R. Landau et al., *CHEMTECH*, **9**, 605. Copyright 1979 American Chemical Society.)

2. Unreacted EB is distilled away in reactor 2, and recycled as shown. Residue (HP) is sent to reactor 3.
3. Excess propylene is added to reactor 3 containing HP and a catalyst to make PO and MBA.
4. The mixture from reactor 3 is sent to reactor 4. The propylene excess is distilled away, and recycled to reactor 3.
5. The mixture is then sent to reactor 5. PO is distilled off. This is product. The still bottoms are sent to reactor 6.
6. EB is separated from reactor 6 and recycled. The remainder of the mixture is sent to reactor 7. The MBA is dehydrated to S.
7. S is purified in reactor 8. The residue from 7 containing acetophenone is hydrogenated in reactor 9 to MBA. This is recycled to reactor 7.

B. *Example 2: A Two-Product Process—Preparation of t-Butanol and Propylene Oxide*

The production equipment is similar to that shown in Figure 4.10. The equation for this process is as follows:

$$\underset{\text{isobutane}}{(CH_3)_3CH} + O_2 \rightarrow \underset{t\text{-butyl hydroperoxide (BHP)}}{(CH_3)_3COOH}$$

$$BHP + \underset{\text{propylene}}{CH_2{=}CHCH_3} \rightarrow PO + \underset{t\text{-butanol}}{(CH_3)_3COH}$$

t-Butanol (TBA) is used as a gasoline octane improver.

Another major use of *t*-butanol involves its conversion to isobutene, a product used for gasoline production.

$$TBA \rightarrow (CH_3)_2C{=}CH_2$$

Landau et al. (1979) list the criteria for success in this type of reaction:

1. The co-product has to have a market value similar to the main product and should have a variety of uses.
2. It should be possible to change the ratio of products by process changes, depending on demand.
3. It should be possible to be able to carry out the PO process with alternate co-products.

The following equation shows an older alternate route to propylene oxide production that is clearly not as efficient.

$$CH_2{=}CHCH_3 + Cl_2 + H_2O \rightarrow$$

$$CH_3CH(OH)CH_2Cl + Ca(OH)_2 \rightarrow CH_3\underset{\diagdown O \diagup}{CHCH_2} + CaCl_2$$

The chlorine added is not used in the product, but ends up as inexpensive calcium chloride, a substance for which there is no need in this quantity.

4.3.3. Example 3: A Two-Product Reaction—Preparation of Phenol and Acetone from Propylene

We now examine another reaction using propylene as a raw material, yielding two products that are on the list of the 50 largest-volume chemicals prepared in the United States.

The equation in Figure 4.11 shows how propylene reacts with benzene to give cumene, which then forms a hydroperoxide with oxygen. The latter product can be decomposed to give phenol and acetone. The mechanism of this rearrangement is shown in Figure 4.12.

4.3.4 Preparation of Butyraldehyde by the *Hydroformylation* Reaction

We now examine the reaction of propylene with synthesis gas—an example of a hydroformylation reaction that is general for all olefins and for formaldehyde

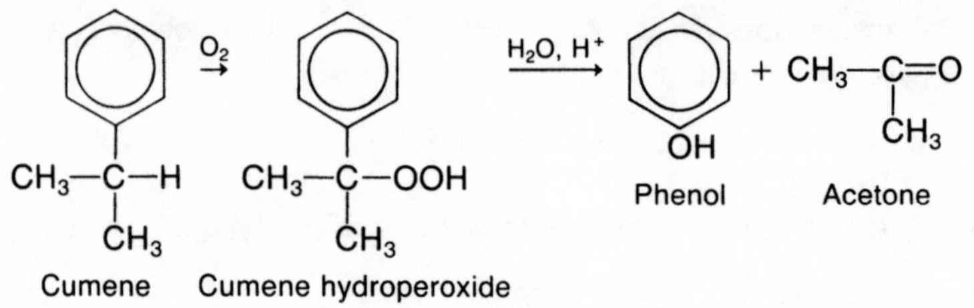

$$C_6H_5\text{–}C(CH_3)_2H \xrightarrow{O_2} C_6H_5\text{–}C(CH_3)_2\text{–}OOH \xrightarrow{H_2O,\ H^+} C_6H_5OH + CH_3\text{–}C(=O)\text{–}CH_3$$

Cumene Cumene hydroperoxide Phenol Acetone

FIGURE 4.11 Equation for the preparation of phenol and acetone. (From R. T. Morrison and R. N. Boyd *Organic Chemistry*, 4th ed., Allyn and Bacon, Inc., Boston, 1983, p. 961.)

(Section 4.2.9.B). The products are mixtures of isomeric aldehydes. Formally, the syngas is added as —H and —HCO to the double bond. Products from both modes of addition are found.

$$CH_3CH{=}CH_2 + CO + H_2 \rightarrow \underset{n\text{-butyraldehyde}}{CH_3CH_2CH_2CHO} + \underset{\text{isobutyraldehyde}}{CH_3CH(CHO)CH_3}$$

Additional products from reaction of the syngas with olefins where the double bond has migrated are often found. The straight-chain aldehyde is usually formed in larger quantities. Reviews of this reaction have been given by Weissermel and Arpe (1978, pp. 112–119), Falbe and Bahrmann (1984, pp. 761–762), and Pruett (1986). The catalyst dicobalt octacarbonyl $Co_2(CO)_8$ is the most commonly used. This material as well as rhodium and ruthenium compounds can be modified with amine- or phosphine-containing ligands. Thus we have considerable flexibility in choice of catalysts. $HCo(CO)_3$, the active catalyst, is soluble in the reaction to give a homogeneous catalyst process. Rhodium, the newer catalyst discussed below, is just as active and can be used at a much lower pressure.

A. Mechanism for the Hydroformylation

Figure 4.12A gives the mechanism of the hydroformylation according to Prins (1980). It is written for the rhodium complex $RhCl(CO)(PPh_3)_2$. It has four steps:

1. Rh catalyst takes up the olefin. See equation A or B.
2. Olefin is inserted into the Rh–H bond. See equation C or D.
3. Alkyl group migrates to a CO ligand. See equation E or F.
4. H_2 undergoes oxidative addition to the complex.

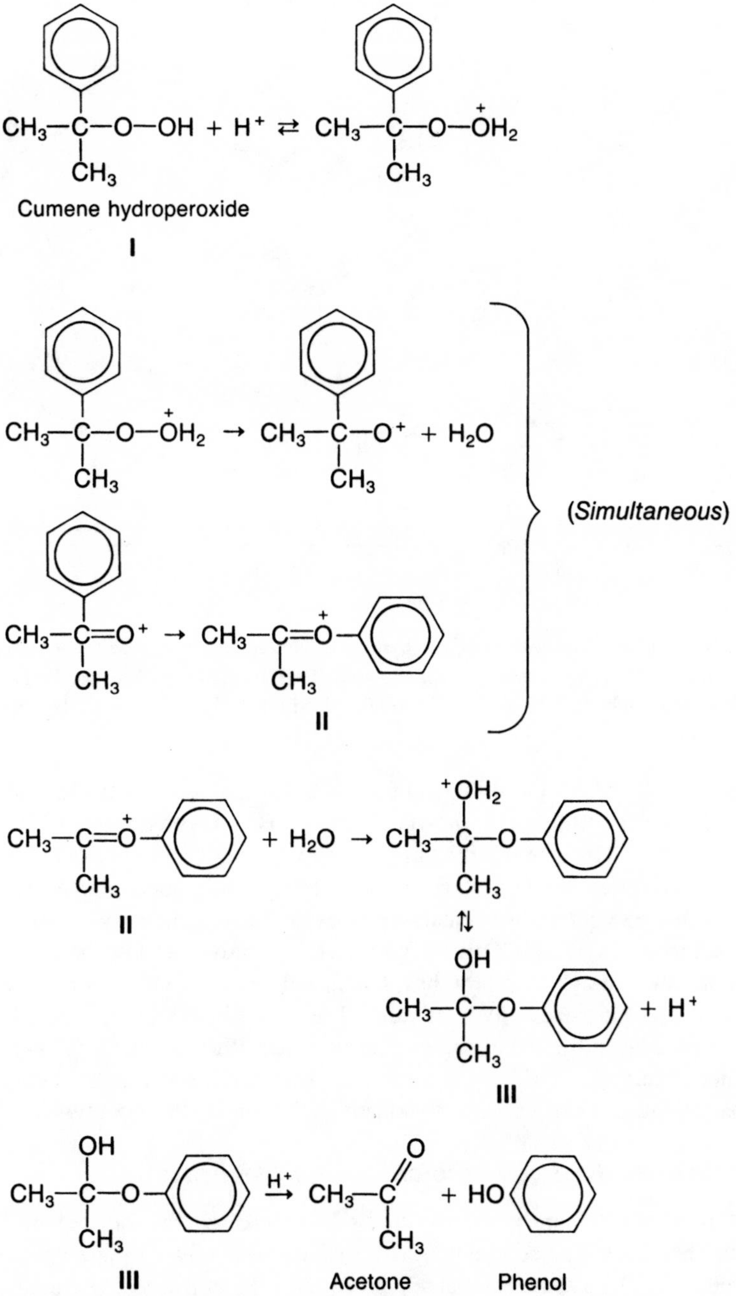

$$CH_3{-}C(C_6H_5)(CH_3){-}O{-}OH + H^+ \rightleftarrows CH_3{-}C(C_6H_5)(CH_3){-}O{-}\overset{+}{O}H_2$$

Cumene hydroperoxide

I

$$CH_3{-}C(C_6H_5)(CH_3){-}O{-}\overset{+}{O}H_2 \rightarrow CH_3{-}C(C_6H_5)(CH_3){-}O^+ + H_2O$$

$$CH_3{-}C(C_6H_5)(CH_3){=}O^+ \rightarrow CH_3{-}C(CH_3){=}\overset{+}{O}{-}C_6H_5$$

II

(*Simultaneous*)

$$\underset{\textbf{II}}{CH_3{-}C(CH_3){=}\overset{+}{O}{-}C_6H_5} + H_2O \rightarrow CH_3{-}C(^+OH_2)(CH_3){-}O{-}C_6H_5$$

$$\rightleftarrows \underset{\textbf{III}}{CH_3{-}C(OH)(CH_3){-}O{-}C_6H_5} + H^+$$

$$\underset{\textbf{III}}{CH_3{-}C(OH)(CH_3){-}O{-}C_6H_5} \xrightarrow{H^+} \underset{\text{Acetone}}{CH_3{-}C(=O){-}CH_3} + \underset{\text{Phenol}}{HO{-}C_6H_5}$$

FIGURE 4.12 Mechanism for the preparation of phenol and acetone from cumene. (From Morrison and Boyd, 4th ed., 1983, pp. 962–963.)

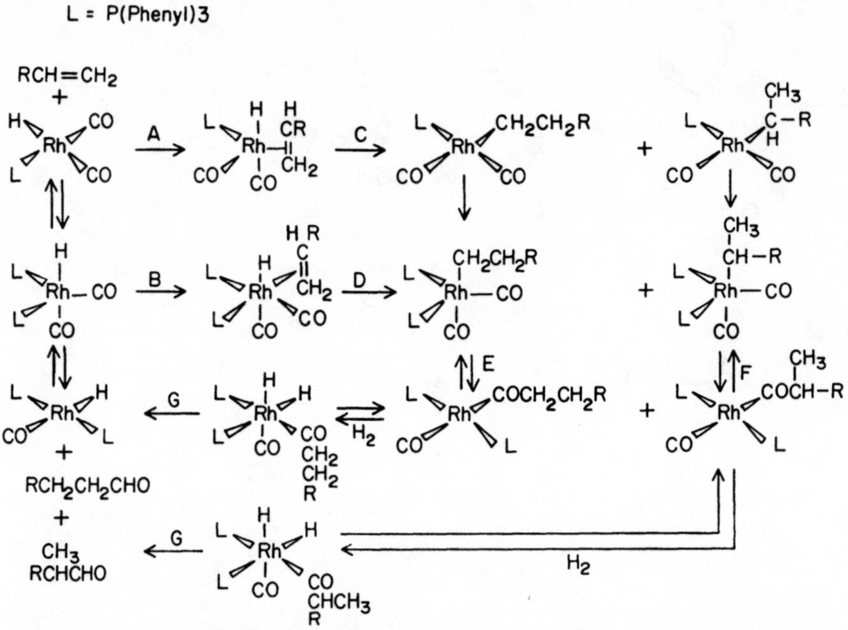

L = P(Phenyl)3

FIGURE 4.12A Mechanism of the hydroformylation reaction. (Reproduced with permission from R. Prins, *Chemistry and Chemical Engineering of Catalytic Processes*, R. Prins and G. Schuit, Eds. Copyright 1980, Martinus Nijhoff/Dr. W. Junk, Dordrecht, The Netherlands, pp. 423–425.)

This is followed by the elimination of a linear and a straight chain aldehyde (G). Two possibilities are shown. The first (B), favored by a high catalyst concentration, is where the rhodium complex adding the olefin has five attached groups. The steric hindrance produces a predominantly linear aldehyde mixture. Too high a concentration of catalyst causes the catalytic activity to drop because of the formation of $Rh(CO)(PPh_3)_3$, which is inactive. At low catalyst concentrations, the catalyst complex has 4 attached groups (A). This gives a lower linear content in the aldehyde mixture. The use of high CO pressures slows the reaction because the catalyst structure becomes $Rh(COR)(CO)_2(PPh_3)_2$ which cannot undergo the oxidative addition of H_2. Rhodium is a better catalyst than cobalt because it carries out the oxidative addition more smoothly.

B. Uses of the Two Hydroformylation Aldehydes

Butyraldehyde can be transformed to *n*-butanol by hydrogenation. This reaction has been studied recently as an example of how to scale up a reaction (Cropley et al., 1984). Butyraldehyde can also be converted to 2-ethylhexanol by the aldol condensation reaction. The latter is the most important alcohol for plasticizers.

Isobutyraldehyde the branched chain aldehyde from the above hydroformylation is a source of many industrially important glycols.

$$\underset{\text{isobutyraldehyde}}{CH_3\underset{\underset{CH_3}{|}}{C}HCHO} + 2HCHO \rightarrow \underset{\text{neopentyl glycol}}{CH_3\overset{\overset{CH_2OH}{|}}{\underset{\underset{CH_3}{|}}{C}}CH_2OH} + HCOOH$$

The preceding reaction is a Cannizzaro reaction occurring with aldehydes having no alpha hydrogen.

Neopentyl glycol is used in unsaturated polyester resins to enhance alkali resistance of the cured product.

4.3.5 Isopropanol

For many years, the industrial procedure for preparing isopropanol from propylene has involved the preparation of the sulfate ester intermediate. Propylene is absorbed in 94% H_2SO_4. The ester is hydrolyzed after diluting to a 40% acid strength.

$$CH_3CH{=}CH_2 + H_2SO_4 \longrightarrow CH_3\underset{\underset{OSO_3H}{|}}{C}HCH_3 \qquad \text{(A)}$$

$$\text{(A)} + H_2O \longrightarrow CH_3\underset{\underset{OH}{|}}{C}HCH_3 + H_2SO_4$$

This process is still in use today. For economic and environmental reasons, the H_2SO_4 has to be recovered for reuse. This is carried out by concentration by distillation of water. The acid has dissolved high boiling by-products in it. These are oxidized by adding some nitric acid during the distillation. The recovery process causes corrosion, and air and water pollution problems.

Many of these problems are overcome in a new procedure by Deutsche Texaco (Weissermel and Arpe, 1978, pp. 175–178; Neier and Woellner, 1973), which uses a reactor filled with a fixed bed of a strongly acidic ion-exchange resin, such as a sulfonated styrene–divinyl benzene copolymer. Liquid water and gaseous propylene are fed to the head of the reactor in a ratio of 12–15 mol water per mol propylene. This ratio suppresses higher boiling by-products. The reaction takes place at 130–160° and 80–100 bar. Isopropanol comes from the reactor as a dilute aqueous solution to be concentrated by distillation. The selectivity is 92–94%, and the catalyst lasts 8 months.

One of the major uses of isopropanol is for the manufacture of acetone. The favored method is dehydrogenation over Cu–Zn catalyst at 500° at 3 bar for a conversion of 98%. Years ago, the source of acetone was the fermentation route (see Section 3.1.5), and some of it reduced to make isopropanol.

4.4 INDUSTRIAL PROCESSES USING THE BUTENE FRACTION OF REFINERY STREAMS

Huge volumes of chemical substances are based on ethylene as a raw material, and somewhat less on propylene. Only one four-carbon chemical, butadiene, is on the list of 50 chemicals produced in highest volume in the United States. Synthetic rubber, vital in World War II, was based on C_4 chemistry. We have discussed ethylene and propylene chemistry, emphasizing the industrial aspects, and will now do the same for butene.

The raw material source for the four-carbon stream is the catalytic cracking of petroleum and the steam cracking of naphtha. Only 5 percent of natural gas is in this chain length, existing as butane. The components of this stream are *n*-butane, butene-1, butene-2, 2-methylpropene (isobutylene) and butadiene.

4.4.1 Synthesis of Butadiene

The following equation shows the three components of the four-carbon stream that are conversed to butadiene by dehydrogenation:

$$CH_3CH_2CH_2CH_3 \quad CH_2{=}CHCH_2CH_3 \quad CH_3CH{=}CHCH_3$$

$$CH_3CH_2CH_2CH_3 \xrightarrow{-2H_2} CH_2{=}CHCH{=}CH_2 \xleftarrow{-H_2} CH_3CH{=}CHCH_3$$

This is the main route to producing this material. It is a thermal dehydrogenation, occurring at 600–700°C, even with catalysts. There are six processes in use, reviewed by Weissermel and Arpe (1978, pp. 97–99). With butane as starting material, the temperature is 130° higher than with the butenes, and coke deposition, isomerization, and polymerization are seen. To reduce this, dehydrogenation in the presence of steam or reduced pressure is carried out. This favorable result occurs because the partial pressure of the hydrocarbon is reduced. With the steam process, the catalyst has to be water stable.

Products made from ethylene can be converted to butadiene as shown in the following equation:

$$CH_3CHO + C_2H_5OH \xrightarrow{TaO,\ SiO_2} CH_2{=}CHCH{=}CH_2$$

4.4.2 Chemicals from Butadiene and Isobutylene

Polymers are the current chief use of C_4 products. Their use is as various rubbers and resins, for example,

Styrene–butadiene rubber
Polybutadiene rubber
Butadiene + acrylonitrile rubber
Acrylonitrile + butadiene + styrene resin (ABS resin)
Butyl rubber

4.4.3 Other Chemicals Produced from Four-Carbon Raw Materials

The following is a list of important industrial products that use one of the butenes as a raw material:

Hexamethylenediamine
t-Butanol
Butylated phenols
Methyl ethyl ketone
Maleic anhydride

A. Hexamethylenediamine

This is important as a raw material for nylon. The annual production for 1984 was 1.2 billion pounds. The synthesis is by reduction of adiponitrile, as follows:

$$\underset{\text{adiponitrile}}{NC(CH_2)_4CN} \xrightarrow{\text{3.5 MPa; 150°C}} \underset{\text{hexamethylenediamine}}{H_2N(CH_2)_6NH_2}$$

The catalysts used are Co and Ni with metal promoters with ammonia to prevent formation of poly amines.

B. Preparation of Adiponitrile

Adiponitrile is another substance that has had many processes developed for it. The justification for spending the time is that there is a large demand for it, and hence the economic necessity for the lowest possible cost. Two methods are briefly described here.

The most economical method for preparing adiponitrile is by addition of hydrogen cyanide to butadiene. The catalyst is nickel phosphine or phosphite with Al or Zn promoters. A commonly used catalyst has the structure $Ni[P(OC_6H_5)_3]_4$.

$$CH_2{=}CHCH{=}CH_2 + 2HCN \rightarrow NC(CH_2)_4CN$$

The mechanism involves 1, 4-addition to butadiene. Then the double bond shifts to the end two carbons, and the second HCN adds in an anti-Markownikoff mode. These two steps are catalyzed by the same catalyst. (See Parshall, 1980, p. 70; Tolman, 1986.)

The second method involves the electrohydrodimerization of acrylonitrile, a Monsanto process. This starts with acrylonitrile and consists of two steps:

$$CH_2{=}CHCN \xrightarrow{2e} \overset{\ominus}{C}H_2\overset{\ominus}{C}HCN$$

$$CH_2{=}CHCN + \overset{\ominus}{C}H_2\overset{\ominus}{C}HCN \longrightarrow NC\overset{\ominus}{C}HCH_2CH_2\overset{\ominus}{C}HCN \xrightarrow{2H^+} NC(CH_2)_4CN$$

Manufacturing Process Details for the Monsanto Adiponitrile Process. (See Weissermel and Arpe, 1978, 218). An electrolysis system is set up where the cathode–anode system contains a two-phase emulsion. The aqueous phase contains a saturated solution of acrylonitrile as its tetraalkylammonium *p*-toluene sulfonate salt. This is where the reaction takes place. The organic phase contains most of the acrylonitrile and the product adiponitrile. As the reaction proceeds, the acrylonitrile in the aqueous phase becomes used up, being converted to adiponitrile which goes to the organic phase. More acrylonitrile goes into the aqueous phase to keep the solution saturated. So, with rapid mixture to permit rapid transfer between phases, the reaction proceeds. The cathode is graphite and the anode is magnetite (Fe_3O_4). The preceding salt coats the cathode, and prevents water electrolysis with hydrogen formation, which would hydrogenate acrylonitrile to propionitrile. This is an example of phase transfer catalysis, discussed in Section 11.1.3.B.

C. Hydrogenation of Adiponitrile

The flow diagram in Figure 4.13 for the reduction of adiponitrile describes a continuous catalytic reduction over a choice of copper–cobalt, cobalt–aluminum, or several other catalysts. The reaction solvent is liquid ammonia. The three ingredients are mixed and heated to 90°C and led to the reactor containing the catalyst. The recycle streams consisting of the excess hydrogen and ammonia are controlled to hold the temperature at 120°C. The reaction mixture is sent to

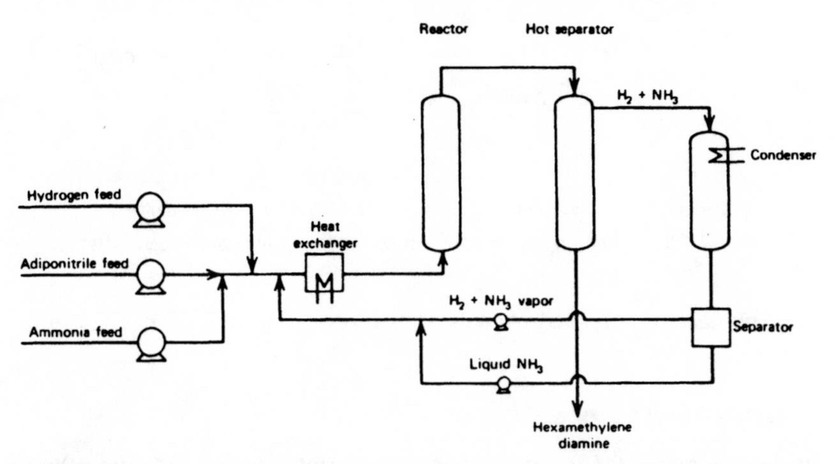

FIGURE 4.13 Hexamethylenediamine by hydrogenation of adiponitrile: flow diagram. (Reprinted with permission from R. L. Sandridge and H. B. Staley, *Encyclopedia of Chemical Technology*, 3rd ed., R. Kirk and D. Othmer, Eds., Wiley, New York, Vol. 2, 363, 1978.)

the hot separator where the hydrogen and ammonia are removed for recycle. See Sandridge and Staley, 1978, for further discussion of this process.

D. t-Butanol

This material is prepared from isobutene by hydration in a similar fashion to ethylene and propylene in this case using dilute mineral acid. Adjustment of conditions yields a dimer, called diisobutylene (2,4,4-trimethylpentene isomers) and a polymer.

$$CH_3C(CH_3){=}CH_2 + H_2O \rightarrow (CH_3)_3COH$$

It has two interesting uses: (1) for preparation of *t*-butyl methyl ether, a compound developed as an antiknock agent for gasoline, (2) in the catalytic oxidation to methacrylic acid, needed for the important monomer methyl methacrylate.

E. Butylated Phenols

Another use for isobutene is in the preparation of the butylated phenols, which are used in a wide variety of substances added for antioxidant properties in plastics, foodstuffs, etc. The important antioxidant BHT is given as an example in Figure 4.14. The reaction shown is carried out as in a Friedel–Crafts reaction.

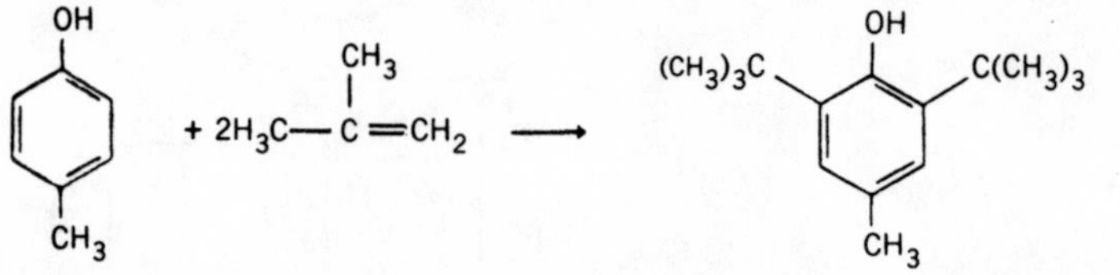

$$p\text{-Cresol} + 2H_3C{-}C(CH_3){=}CH_2 \text{ (Isobutene)} \longrightarrow \text{Butylated hydroxytoluene (BHT)}$$

p-Cresol Isobutene Butylated hydroxytoluene (BHT)

FIGURE 4.14 Equation for preparation of butylated hydroxytoluene (BHT).

A large number of butylated phenols with various substitutions have been developed.

F. *Methyl Ethyl Ketone*

Methyl ethyl ketone (MEK) is an important solvent for vinyl and nitrocellulose lacquers and acrylic resins. 1-Butene and 2-butene mixtures can be hydrated using concentrated sulfuric acid to yield 2-butanol. This can be oxidized to MEK over a zinc oxide or zinc–copper catalyst.

$$\left.\begin{matrix} CH_2{=}CHCH_2CH_3 \\ CH_3CH{=}CHCH_3 \end{matrix}\right\} \rightarrow CH_3CH_2CH(OH)CH_3 \rightarrow CH_3CH_2C({=}O)CH_3$$

G. *Maleic Anhydride*

See Section 4.7.2.A for the discussion of the synthesis of this material from a mixture of butenes, using air oxidation with a V_2O_5 catalyst.

4.5 INDUSTRIAL CHEMICALS FROM BENZENE

There are two sources of benzene. The first is catalytic reforming of light or heavy naphtha, which was mentioned in Section 4.1.3.F. A typical analysis of the product of such a process is as follows:

Benzene, 10–15%
Xylene, 35–45%
Toluene, 50%

As will be seen from this text, there is a larger industrial demand for benzene and xylene than for toluene. To remedy this situation, a process was developed for dealkylation of toluene to benzene by hydrogenation.

$$H_2 + CH_3C_6H_5 \rightarrow C_6H_6 + CH_4$$

More than one-half the toluene produced in the United States is converted to benzene in this way. The conditions used are 500–650°C using Cr, Mo, or Co oxides on alumina supports.

The following chemicals are made from benzene as a raw material:

Adipic acid
Caprolactam
Aniline and other benzene derivatives
Styrene (see Section 4.2.10)
Phenol (see Section 4.3.4)
Maleic anhydride

4.5.1 Adipic Acid and Caprolactam

Adipic acid is used, along with hexamethylene diamine, for the manufacture of nylon 6,6. Caprolactam is the raw material for nylon 6. These nylon structures are shown in Figure 4.15. Figure 4.16 shows how both these nylon monomers

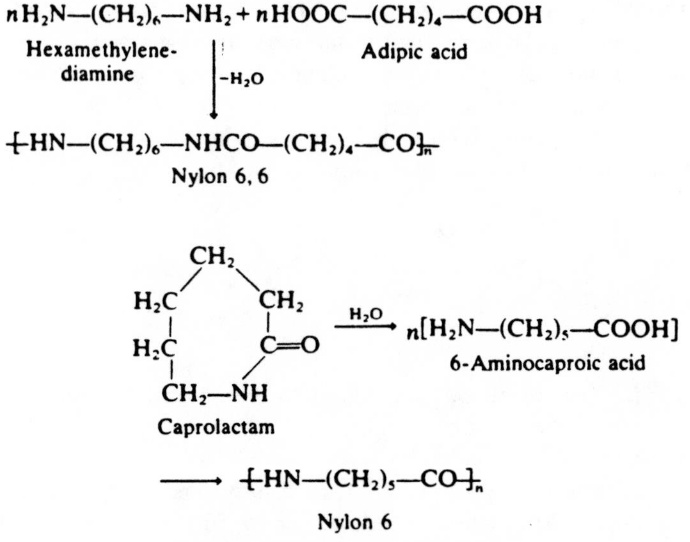

$$nH_2N\text{—}(CH_2)_6\text{—}NH_2 + nHOOC\text{—}(CH_2)_4\text{—}COOH$$

Hexamethylene-diamine; Adipic acid

$$\xrightarrow{-H_2O} \;\; +\!\!\!\!\!\!\!\quad HN\text{—}(CH_2)_6\text{—}NHCO\text{—}(CH_2)_4\text{—}CO\,\text{}_n$$

Nylon 6, 6

Caprolactam (ring: CH_2—CH_2—CH_2—CH_2—CH_2—NH—C=O) $\xrightarrow{H_2O}$ $n[H_2N\text{—}(CH_2)_5\text{—}COOH]$ 6-Aminocaproic acid

$$\longrightarrow \; [HN\text{—}(CH_2)_5\text{—}CO]_n$$

Nylon 6

FIGURE 4.15 Nylon structures.

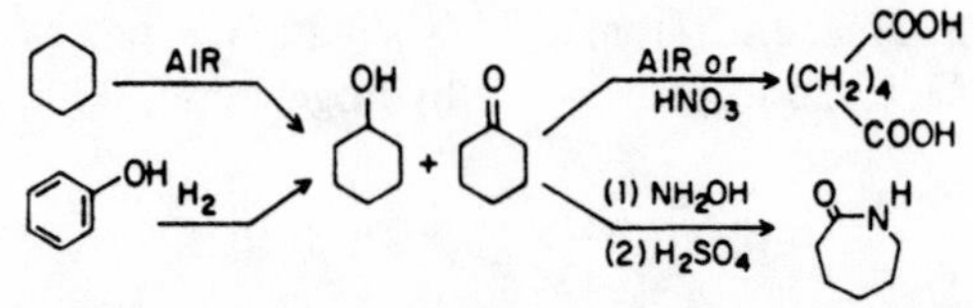

FIGURE 4.16 Reaction scheme for the production of adipic acid and caprolactam from "mixed oil," and the raw material choices for this intermediate. (Reprinted with permission from G. W. Parshall, *Homogeneous Catalysis*, Wiley, New York, 1980, p. 191.)

are formed from the same raw material mix, a mixture of cyclohexanone and cyclohexanol. This mixture is called "mixed oil." As shown in Figure 4.16, (Parshall, 1980, p. 191) there are two commercial methods for preparing mixed oil, air oxidation of cyclohexane and hydrogenation of phenol. The latter process gives cyclohexanone or cyclohexanol by adjustment of the catalyst. The catalyst used is palladium on carbon or alumina. This industrial process is carried out in slurry or fixed-bed systems.

Cyclohexane can also be directly oxidized to adipic acid but the yield has never been as high as when the process is used where the cyclohexanol–cyclohexanone mixture is isolated.

The conversion of cyclohexane to mixed oil is brought about by cobalt or manganese naphthenates or benzoyl peroxide at 125–165°C and 8–15 bar with air or oxygen. This is considered to proceed to the hydroperoxide stage as shown in the equation in Figure 4.17 (Parshall, 1980, p. 189), where benzoyl peroxide is the initiator. The cyclohexyl radical $R_2CH\cdot$ combines with oxygen to form a cyclohexyl peroxy radical. A small quantity of this abstracts a hydrogen from a cyclohexane molecule, yielding a cyclohexyl hydroperoxide molecule and a cyclohexyl radical for another cycle.

The cyclohexyl peroxy radical decomposes to cyclohexanol and cyclohexanone as shown in the following equation:

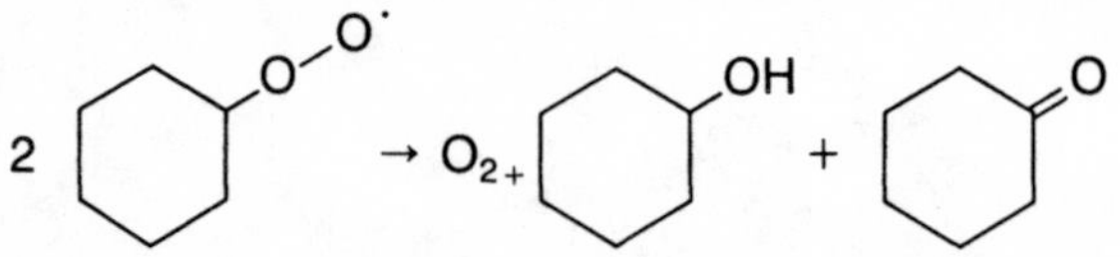

$$2\,C_6H_{11}OO\cdot \rightarrow O_2 + C_6H_{11}OH + C_6H_{10}O$$

Adipic Acid. The mixed oil is oxidized to adipic acid by nitric acid. The nitrogen oxides are reoxidized to nitric acid by air and so oxygen is the true oxidizing agent.

The manufacturing process in brief consists in feeding the mixed oil to a mixture of copper(II) nitrate and NH_4VO_3 in 45–50% nitric acid at 70–80°C. The evolved nitrogen oxides are recycled to a nitric acid manufacturing unit for

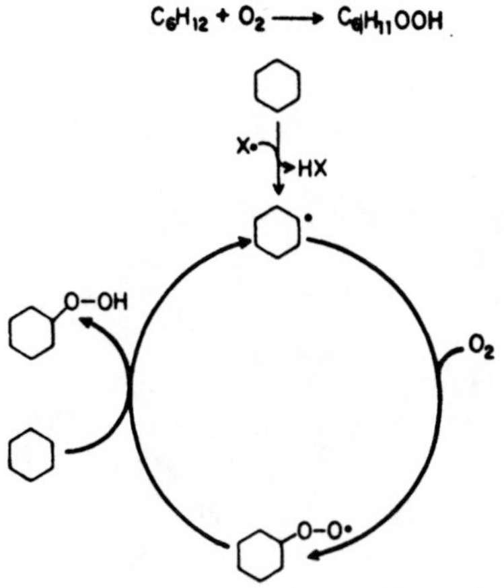

$$C_6H_{12} + O_2 \longrightarrow C_6H_{11}OOH$$

FIGURE 4.17 Cycle for the oxidation of cyclohexane to cyclohexyl hydroperoxide. (Reprinted with permission from G. W. Parshall, *Homogeneous Catalysis*, Wiley, New York, 1980, p. 189.)

reoxidation to nitric acid. See Parshall (1980, p. 193) and Figure 4.18. This process has been described in Section 2.5.1.

Caprolactam. Liquid-phase catalytic hydrogenation of phenol over a Pd catalyst yields cyclohexanone (for caprolactam), or cyclohexanol (for adipic acid) in high yield (with a Ni catalyst) (Danby and Campbell, 1978). The former is reacted with hydroxylamine sulfate to give cyclohexanone oxime. The process has been developed to give a molten oxime layer and a saturated ammonium sulfate layer:

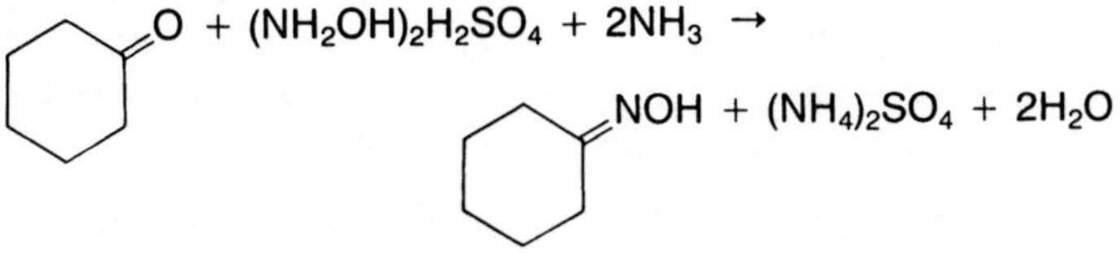

$$C_6H_{10}{=}O + (NH_2OH)_2H_2SO_4 + 2NH_3 \rightarrow C_6H_{10}{=}NOH + (NH_4)_2SO_4 + 2H_2O$$

The oxime is quantitatively converted to caprolactam by reaction with oleum. On neutralization with ammonia, two layers result, a product layer and an

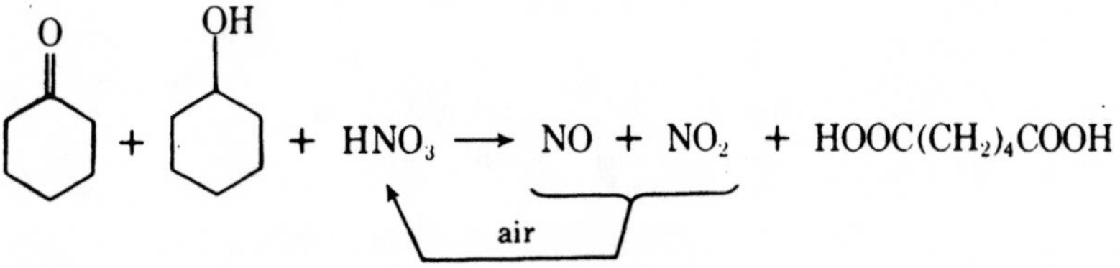

$$C_6H_{10}O + C_6H_{11}OH + HNO_3 \longrightarrow NO + NO_2 + HOOC(CH_2)_4COOH$$

FIGURE 4.18 Conversion of "mixed oil" to adipic acid using nitric acid. (Reprinted with permission from G. W. Parshall, *Homogeneous Catalysis*, Wiley, New York, 1980, p. 193.)

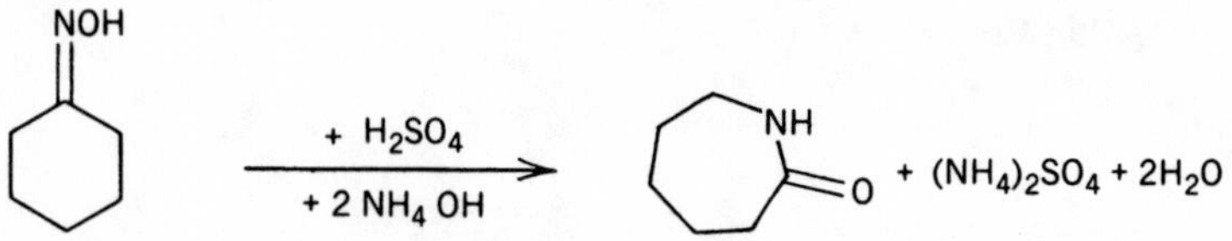

$$\text{cyclohexanone oxime (NOH)} \xrightarrow[+\ 2\ NH_4OH]{+\ H_2SO_4} \text{caprolactam} + (NH_4)_2SO_4 + 2H_2O$$

FIGURE 4.19 Equation for conversion of cyclohexanone oxime to caprolactam.

$(NH_4)_2SO_4$ layer. The equation is shown in Fig. 4.19 (Fisher and Crescentini, 1982).

A process that gives less ammonium sulfate by-product involves the conversion of benzoic acid to cyclohexanecarboxylic acid by hydrogenation with Pd on carbon at 170°C. The latter is treated with nitrosylsulfuric acid to yield caprolactam directly (Rylander, 1983, pp. 9–10). See Figure 4.20.

Ammonium sulfate presents a disposal problem, over and above the amount used as fertilizer. To overcome this, oleum is replaced by phosphoric acid. Ammonium phosphate is more useful as a fertilizer than ammonium sulfate because it has both phosphorus and nitrogen in usable form.

The polymerization of hexamethylenediamine with adipic acid to give nylon 6,6 and of caprolactam to give nylon 6 is described by Putscher (1982).

4.5.2 Aniline and Other Benzene Derivatives

Table 4.3 gives the industrial conditions for making various benzene derivatives. The conversion of benzene to aniline is an excellent example of a reaction that has been carried out in different ways because of environmental concerns as well as economic pressures. The most traditional raw material is nitrobenzene. In recent years chlorobenzene, phenol, and even benzene itself have been used. We now discuss five methods for preparing aniline.

1. For the conversion of nitrobenzene to aniline, using iron turnings and hydrochloric acid was the method of choice until recent times. Iron oxide pigments were obtained from the sludge. When demand for aniline increased, far too much iron was produced for the requirements of the dyestuffs industry. This led to the development of catalytic hydrogenation processes.

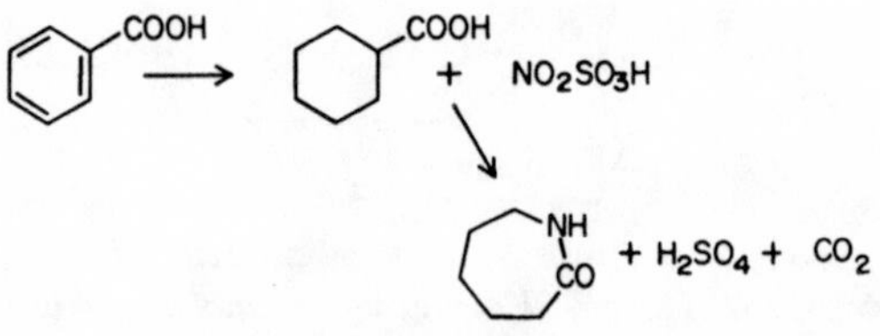

$$C_6H_5COOH \longrightarrow C_6H_{11}COOH + NO_2SO_3H \longrightarrow \text{caprolactam (NH, CO)} + H_2SO_4 + CO_2$$

FIGURE 4.20 Equation for conversion of benzoic acid to caprolactam via cyclohexanecarboxylic acid.

TABLE 4.3 Industrial Reaction Conditions for Benzene Derivatives

Derivative	Temperature (°C)	Pressure (bar)	Catalyst (psig)	Yield (%)
Ethyl benzene[a]	40–100	30–100	$AlCl_3(HCl)$	100
Styrene[b]	600–700	Atmos	Fe + Cr oxides	90
Cumene[c]	200–250	400–600	H_3PO_4 kieselguhr	96–97
Dodecyl benzene[d]	200–240	15–25	H_3PO_4 or other acids	
Chlorobenzene[e]	80–100	Atmos	$FeCl_3$	81[e] (mono)
Nitrobenzene[f]	55	Liq	H_2SO_4	95–98
Aniline[g]	210–220	880–1130	Cu salts	96
Aniline[h]	250–270	20	Cu on silica	95

[a]Reactants: benzene and ethylene + catalyst.
[b]Reactants: ethyl benzene + catalyst.
[c]Reactants: benzene and propylene + catalyst.
[d]Reactants: benzene and dodecene + catalyst.
[e]Reactants: benzene and chlorine + catalyst. Product also made with benzene, oxygen, and HCl.
[f]Reactants: benzene, concentrated HNO_3, + catalyst.
[g]Reactants: chlorobenzene, NH_3, + catalyst.
[h]Reactants: nitrobenzene, hydrogen, + catalyst.

2. Catalytic hydrogenation of nitrobenzene using a fixed bed of nickel sulfide is carried out at 375°C. The aniline selectivity is 99%. The catalyst can be regenerated.
3. Using a fluidized bed of Cu, Cr, Ba, or Zn on SiO_2 at 270–290°C and 1–5 bar, nitrobenzene can be hydrogenated at 99.5% selectivity.
4. Chlorobenzene and phenol can be converted to aniline by ammonolysis. The chlorobenzene reaction uses catalysis by copper(I) chloride. Ammonolysis of phenol is carried out in adiabatic fixed bed reactors with catalysis with oxides of Al and Si.
5. Benzene can be converted to aniline using ammonia with a nickel oxide–nickel catalyst containing zirconium promoters. The conditions are 350°C and 300 bar. The selectivity is 97%, and the conversion is 13%.

$$C_6H_6 + NH_3 \rightarrow C_6H_5NH_2 + H_2$$

The catalyst is reduced by the hydrogen and has to be reoxidized.

4.5.3 Maleic Anhydride

See Section 4.7.2.A for a discussion of the synthesis of this material from benzene using air oxidation with a V_2O_5 catalyst.

4.6 INDUSTRIAL CHEMICALS FROM TOLUENE

The source of the raw material toluene was mentioned in Section 4.1.3.F. It is prepared chiefly from the reforming reaction along with benzene and the xylenes. We now discuss four reactions:

1. Benzene and methane by dealkylation
2. Benzene and xylenes by disproportionation
3. Styrene by metathesis
4. Toluene diisocyanate, a major product

The first three reactions were devised to convert the excess toluene to other aromatic raw materials.

1. The dealkylation of toluene to benzene and methane resembles the cracking process discussed in Section 4.1.3.B. It is mentioned in Section 4.5.

$$CH_3C_6H_5 + H_2 \rightarrow C_6H_6 + CH_4$$

2. The commercial process using the disproportionation reaction to the greatest extent is the treatment of toluene to give a mixture of three xylenes and benzene. The reaction is carried out by heating over aluminum trichloride with HCl or other halogenated compounds to 80–125°C and 35–70 bar.

$$CH_3C_6H_5 \rightarrow C_6H_6 + C_6H_4(CH_3)_2 \quad \text{(mix of } o\text{-, } m\text{-, and } p\text{-xylenes)}$$

3. The metathesis reaction has been discussed in Section 4.1.3.G. A promising attempt to make styrene from toluene has been reviewed by Innes and Swift (1981, p. 244). This method involves the preparation of stilbene by a catalytic oxidative coupling over $PbO\text{–}MgAl_2O_4$ at 600°C. Stilbene undergoes a metathesis reaction where it is reacted with ethylene as shown in the equation at 500°C over the oxides of Ca, W, and Si. (See Figure 4.21.) Rylander (1983, pp. 32–33) and Mol (1975, 1983) review several other promising processes. See Table 11.2.

4. The preceding reactions were specifically developed to use up toluene. The reaction discussed here was designed to make a desired product, toluene diisocyanate (TDI). This substance is widely used in polyurethane resins. The equation is shown in Fig. 4.22. 2,4-dinitrotoluene (from nitration of toluene) is

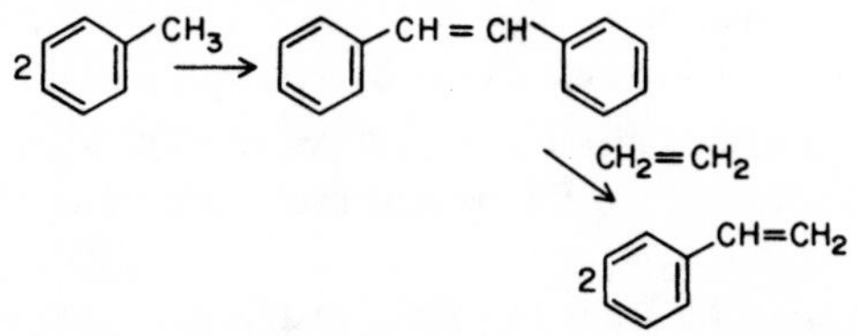

FIGURE 4.21 Reaction scheme for the conversion of toluene to styrene.

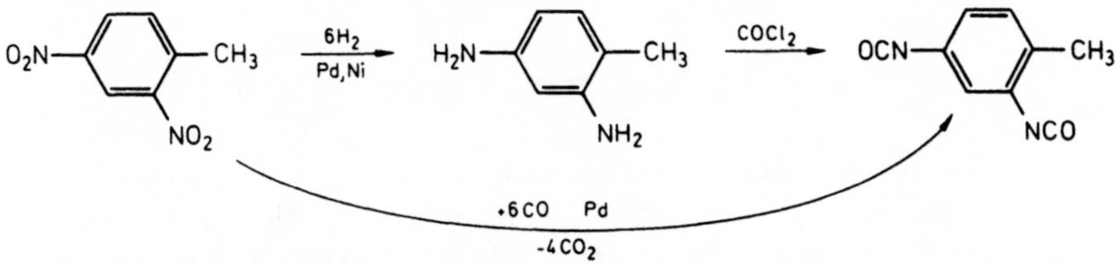

FIGURE 4.22 Equation for preparation of toluene diisocyanate (TDI). From P. Rylander, *Catalysis Science and Technology*, J. R. Anderson and M. Boudart, Eds., Vol. 4, Springer-Verlag, Heidelberg, 1983, p.7.)

reduced to the amines. The reaction is carried out at 400°C and 7 MPa on Raney nickel or 5% palladium as catalyst. The resulting diamine is reacted with phosgene ($COCl_2$) to give the diisocyanate. $COCl_2$ is made from CO and Cl_2 (Hardy, 1982).

An alternate production method is to do the reduction and carbonylation in one step, as shown in Fig. 4.22 (Rylander, 1983, p. 7). In this case the CO in the reducing agent as well as the reactant.

4.7 INDUSTRIAL CHEMICALS FROM XYLENES

The source of the raw material xylene has been mentioned in the discussions on benzene and toluene (see Sections 4.5 and 4.6). The most important use of *p*-xylene is in the preparation of terephthalic acid.

4.7.1 Terephthalic Acid

Several methods for preparing this material follow. This material is used in the preparation of polyester fiber. The U.S. production in 1984 was 6.05 billion pounds.

$$\underset{p\text{-xylene}}{CH_3C_6H_4CH_3} \rightarrow \underset{\text{terephthalic acid}}{HOOCC_6H_4COOH}$$

The main synthesis, called the Amoco process, uses air with catalysts at 190–205°C, and 15–30 bar. The conversion is 95%. The catalyst is copper or manganese naphthenate, a homogeneous catalytic system. The reaction is chemically related to the preparation of the mixture of cyclohexanol–cyclohexanone used in nylon manufacture.

Isophthalic acid is made from *m*-xylene, using the same catalyst mixture. This product is used in alkyd resins.

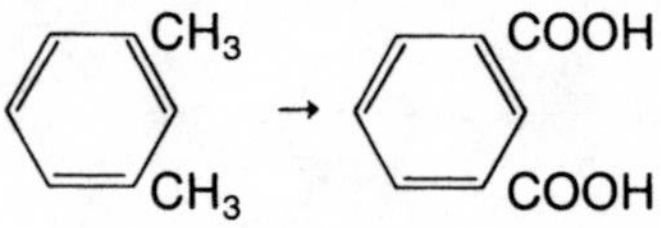

CH_3 / CH_3 → COOH / COOH

Another synthesis of terephthalic acid is by an ammoxidation process, of *p*-xylene as shown in Fig. 4.23 (Rylander, 1983, p. 29). No aldehydic impurities are formed. These are found as by-products in oxidation processes such as the preceding one. See Section 4.3.1.B for another ammoxidation reaction. The intermediate dinitriles can, of course, be hydrogenated to the corresponding diamines.

4.7.2 Phthalic Anhydride

Ortho-xylene is a raw material for phthalic anhydride, one of the 50 largest volume chemicals produced. There is another material—naphthalene—which can be used for the same synthesis. The synthesis procedure is similar for both, namely, air oxidation with V_2O_5 catalysis:

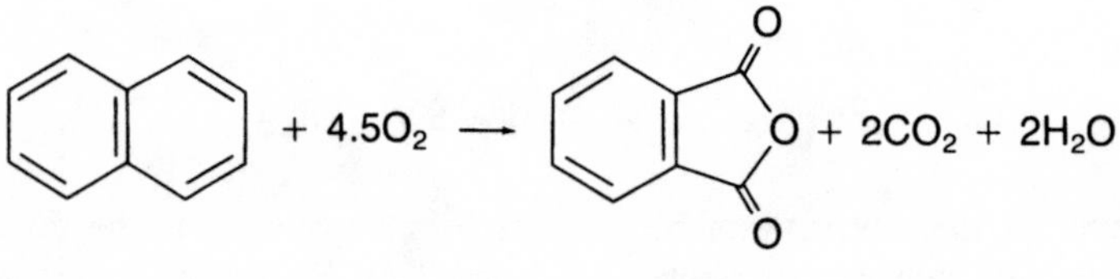

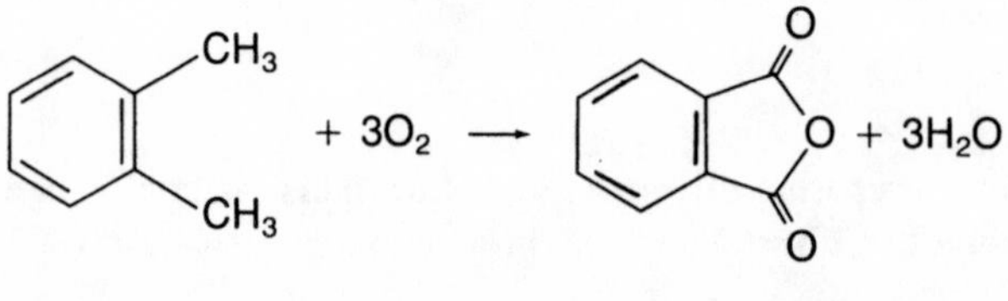

$$\text{naphthalene} + 4.5O_2 \longrightarrow \text{phthalic anhydride} + 2CO_2 + 2H_2O$$

$$o\text{-xylene} + 3O_2 \longrightarrow \text{phthalic anhydride} + 3H_2O$$

The selectivity is 86–91% for naphthalene and 78% for *o*-xylene. This factor favors naphthalene; however, the equations show a considerably smaller oxygen requirement for *o*-xylene and consequently a smaller heat of reaction: $\Delta H =$

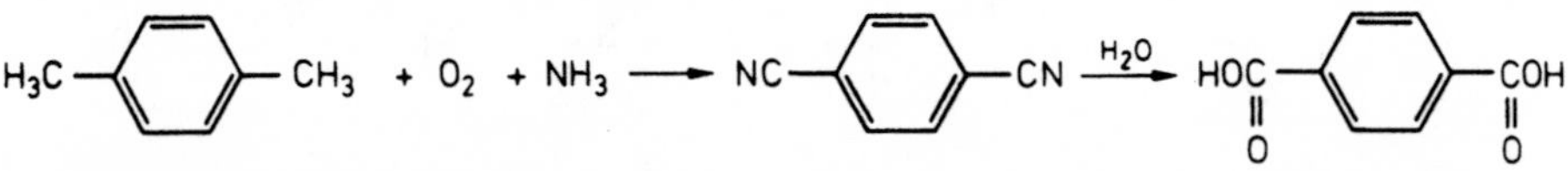

$$H_3C-C_6H_4-CH_3 + O_2 + NH_3 \longrightarrow NC-C_6H_4-CN \xrightarrow{H_2O} HOC(=O)-C_6H_4-C(=O)OH$$

FIGURE 4.23 Equations for terephthalic acid manufacture by ammoxidation. (From P. Rylander, *Catalysis Science and Technology*, J. R. Anderson and M. Boudart, Eds., Vol. 4, Springer-Verlag, Heidelberg, 1983, p. 29.)

−265 kcal compared to −428 kcal for naphthalene, making the *o*-xylene reaction easier to control. Plants are now in operation which can use either raw material depending on availability.

A. Maleic Anhydride

The procedure for this material, widely used in polyesters and paint raw materials, is the same as for phthalic anhydride, that is, the use of air oxidation over V_2O_5 with two alternate starting materials. These are benzene or butenes, either from the cracking of naphtha to obtain ethylene or from the preparation of butadiene from butane.

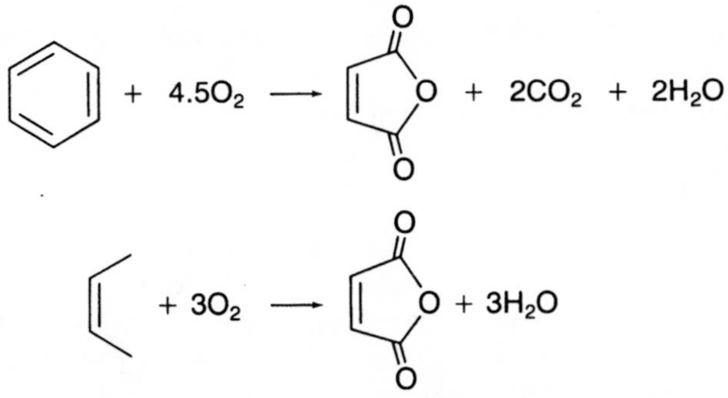

$$C_6H_6 + 4.5O_2 \longrightarrow C_4H_2O_3 \text{ (maleic anhydride)} + 2CO_2 + 2H_2O$$

$$C_4H_8 \text{ (butene)} + 3O_2 \longrightarrow C_4H_2O_3 \text{ (maleic anhydride)} + 3H_2O$$

The ΔH is much lower for the butenes, but this is outweighed by the fact that there is much isobutene present, which is fully oxidized in the process, generating much heat.

The selectivities of these processes are 95% for benzene and 50–60% for the butenes. This, of course, favors the process using benzene, but the following factors should be considered. The benzene process gives 50–60% as the anhydride and the remainder as maleic acid. This is readily dehydrated to the anhydride in a separate step. There is little or no maleic acid in the butene process mother liquor. Another factor to be noted is that in recent years the cost of benzene compared to the butenes has varied considerably with supply and demand.

Another source of maleic anhydride is the mother liquors of the phthalic anhydride process using the naphthalene process.

4.8 INDUSTRIAL CHEMICALS FROM METHANE

We have now discussed six of the seven chemical substances derived from petroleum. From these petrochemicals are derived 90% of all commercially prepared organic substances. The seventh is methane, which we have encountered four times before in this book: First, there was the synthesis of ammonia from nitrogen and hydrogen, the latter derived from methane and steam by the reaction forming synthesis gas, a carbon monoxide–hydrogen mixture (see Section 2.3.1.A). Second, we discussed the significant quantity of methane available from anaerobic digestion of sewage sludge (see Section 3.2.4.B). Third, we discussed the hydroformylation reaction, addition of methane-derived synthesis gas to olefins to yield aldehyde mixtures (see Section 4.3.3). Fourth, the preparation of methanol from syngas was discussed in detail in Section 4.2.5.

We now discuss various large-scale products that use methanol in their manufacture. The approach is that of Frank (1982), who discussed these products under two headings: (1) traditional uses of methanol, and (2) new and evolving uses of methanol. We first briefly review the traditional uses:

4.8.1 Traditional Uses of Methanol

A. Formaldehyde

The manufacture of formaldehyde is carried out by removal of hydrogen from methanol by a catalytic process. This process was referred to in the discussion of ethylene glycol in Section 4.2.9.A. The most up-to-date process involves the use of silver needles or gauzes at 600°C. The methanol is exposed for about 0.01 second, and very little methanol is left. Air is added to the mixture to combine with the hydrogen. This liberates heat, which is needed for the endothermic dehydrogenation step.

This silver catalysis reaction resembles the ethylene oxide synthesis in that the metal significantly reacts only when it is covered with a layer of oxygen. The mechanism is shown in these two equations:

$$CH_3OH_{(g)} + O_{(ads)} \xrightarrow[600^\circ C]{Ag} CH_3O_{(ads)} + OH_{(ads)}$$

$$CH_3O_{(ads)} \rightarrow CH_2O_{(g)} + H_{(ads)}$$

For a further review of this topic, see Rylander (1983, pp. 19–20).

B. *Dimethyl Terephthalate (DMT)*

Polyester fiber is made from the raw materials dimethyl terephthalate and ethylene glycol

$$nCH_3OOC{-}C_6H_4{-}COOCH_3 + nHOCH_2CH_2OH$$
$$\downarrow$$
$$-OOC{-}C_6H_4{-}COOCH_2CH_2{-} \; + 2nCH_3OH$$

The process uses *p*-xylene as a raw material and is a catalyzed air oxidation. The most recent approach is to carry out the oxidation to the acid and esterification in one reactor. The oxidation proceeds in two steps. The conversion of the second methyl group is difficult, but made easier by the esterification of the carboxyl group in *p*-toluic acid. In this continuous process *p*-xylene and partly oxidized product are introduced at the top of the column, and air and methanol at the bottom. The catalyst is a Cu–Mn salt of an organic acid, the temperature is 100–200°C, the pressure is 5–20 bar, and the residence time is 22 hours.

$$CH_3{-}C_6H_4{-}CH_3 \xrightarrow[\text{cat}\; CH_3OH]{\text{air}} CH_3{-}C_6H_4{-}COOCH_3 \longrightarrow CH_3OOC{-}C_6H_4{-}COOCH_3$$

The use of dimethyl terephthalate in this process is gradually being displaced by use of terephthalic acid. Further background on this process is given in Weissemel and Arpe (1978, pp. 341–44).

C. *Methylamines*

Methylamines are prepared from methanol and ammonia at 250–500°C. A laboratory-trained chemist would not be able to predict that this reaction would proceed to a good yield using these raw materials. Methyl chloride would be chosen. However, by using the appropriate catalyst (e.g., aluminum silicate at 10–200 bar), very high selectivity is obtained.

$$CH_3OH + NH_3 \rightarrow CH_3NH_2 + H_2O$$

The mono- and dialkylation products are favored by excess ammonia and by addition of trimethylamine. The workup consists of a pressure and an extractive distillation. For the latter technique, see Section 9.1.1 and Weissermel and Arpe, (1978, p. 45).

The use of methylamines to aminate long chain alcohols is discussed by Balker et al. (1983).

D. Methyl Methacrylate (MMA)

Methacrylic acid esters have many uses, such as polymers for viscosity index improvers and pour point depressants for all-season motor oils. The methyl ester is the usual starting material. In recent years several syntheses have been considered as alternates to the following reaction, but none has taken over completely. Production via this synthesis is still significant.

$$CH_3COCH_3 + HCN \rightarrow (CH_3)_2C(OH)CN \rightarrow CH_3C(CN){=}CH_2 \quad (A)$$

$$(A) + CH_3OH \rightarrow CH_2{=}C(CH_3)COOCH_3$$

E. Acetic Acid

This process, discussed in section 4.2.4.B., is now a major use of methanol.

$$CO + CH_3OH \rightarrow CH_3COOH$$

4.8.2 New and Evolving Uses of Methanol

A. Fuel Uses

1. Methanol is now being used as a fuel either as an up to 10% blend with gasoline or as the neat material.
2. Methyl *t*-butyl ether is being blended into gasoline to meet octane requirements (Taniguchi and Johnson, 1979). See Section 4.4.3.D.
3. Methanol is being used as a feedstock for synthetic gasoline. This is the Mobil MTG process. Methanol is prepared in high selectivity from synthesis gas. Although used in fuel blends, it has only one-half the energy density of gasoline, and so the conversion of methanol to gasoline has been studied. The Mobil MTG (methanol to gasoline) process has successfully done this over Zeolite ZSM-5 catalyst. The product does not have the broad spectrum of product molecular weights of the Fischer–Tropoch synthesis (see Thompson, 1981). (Inui et al, 1983).

B. Nonfuel Uses

Methanol is being used as a chemical feedstock for ethanol (Francoisse and Thyrion, 1983) (catalyst-promoted Co carbonyl):

$$CH_3OH + CO + 2H_2 \rightarrow C_2H_5OH + H_2O$$

ethylene:

$$C_2H_5OH \rightarrow CH_2{=}CH_2 + H_2O$$

and ethylene glycol. For this process see Section 4.2.9.

C. Sewage Treatment

In tertiary treatment of sewage, denitrification is brought about by bacteria that reduce NO_x to N_2. Methanol or another reducing agent must be present.

D. Single-Cell Protein Manufacture (SCP)

Methanol (as well as higher alkanes, higher alcohols, and methane) can be used as a source of carbon for SCP synthesis (Weissermel and Arpe, 1978). Krieger (1980) reports that Phillips Petroleum has perfected a process whereby yeast and bacteria can synthesize proteins from the preceding carbon sources in the presence of aqueous nutrient solutions containing the essential sulfur, phosphorus, and nitrogen compounds. While gaining energy, microorganisms (e.g., Candida yeast) decompose the paraffins step by step to carbon dioxide and at the same time produce rich cellular substances roughly equivalent to fish meal. Of the sources mentioned above, it is best to start with methanol because of its biologically lower oxygen requirements.

4.8.3 Other Uses of Methane

A. Gasoline by the Fischer–Tropsch Synthesis

Among the many products from synthesis gas we now refer to the Fischer–Tropsch synthesis, where a 2:1 mixture of hydrogen–carbon monoxide can be converted into gasoline-sized molecules in the C_8 range. These would be a mixture of olefins and paraffins. The catalysts are various iron salts. M.E. Dry (1982), the supervisor of the only commercial plant now in operation, reviews this process.

$$8CO + 16H_2 \rightarrow C_8H_{16}$$

B. Preparation of Hydrogen Cyanide

There are two methods for preparing hydrogen cyanide. The first is by the ammoxidation of methane. This method, called the Andrussow process, involves heating a mixture of air, ammonia, and methane to 1000–1200°C and leading the hot mixture over a platinum–rhodium catalyst at 20–30 psi. There is a rapid reaction and hydrogen cyanide is formed. For another ammoxidation reaction, see Section 4.3.1.B.

$$CH_4 + NH_3 + 1.5O_2 \rightarrow HCN + 3H_2O$$

This is an exothermic reaction. A review of the details of this process is given by Farmer (1977). Adding hydrogen sulfide to this process increases the effectiveness of the catalyst and prevents deactivation due to coke (Hillebrand, 1984).

The other method for obtaining hydrogen cyanide is as a by-product of the preparation of acrylonitrile from propylene by ammoxidation. For a review of this reaction and the quantity of HCN available, see Section 4.3.1.B.

There are many uses of hydrogen cyanide, the most common being those based on cyanuric chloride. This material is an intermediate that leads to a wide variety of products with many diverse uses.

Preparation of Cyanuric Chloride—a two-step process:

1. Preparation of Cyanogen Chloride

$$HCN + Cl_2 \rightarrow ClCN + HCl$$

The reaction can also be carried out by the electrolysis of aqueous solutions of HCN and ammonium chloride (U.S. Patent 3,294,657) and also by the gaseous oxidation of HCN with oxygen and hydrogen chloride with transition metal catalysts (Jap. Patent 7,036,204).

2. Preparation of cyanuric chloride

$$3ClCN \rightarrow$$

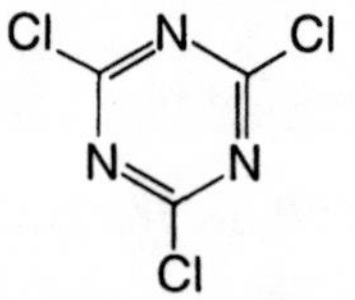

Cl N Cl
N N
Cl

This reaction is carried out by passing the cyanogen chloride through a tube of carbon catalyst.

In 1984, a significant fraction of hydrogen cyanide made in the United States was converted to cyanuric chloride. Triazine herbicides (see Section 10.2.1) and optical bleaches are prepared from cyanuric chloride by Ciba-Geigy.

C. *Preparation of Chlorinated Methanes*

The approach used in preparing chloromethanes is similar to that for the preparation of vinyl chloride—a combination of direct chlorination and oxychlorination. The reaction called oxychlorination used in the preparation of vinyl chloride (see Section 4.2.2) is used to prepare a mixture of chloromethanes. Methane, oxygen, and HCl are used with a copper (II) chloride catalyst to carry this out.

$$CH_4 + HCl + O_2 \rightarrow CH_3Cl + \text{higher chloromethanes} + H_2O$$

The equation for the direct chlorination is $CH_4 + Cl_2 \rightarrow CH_3Cl$ + higher chloromethanes + HCl. The HCl is recycled to the oxychlorination.

D. Preparation of Formic Acid

Formic acid is used for the pH control of dyes and pickling baths. The aluminum and sodium salts are used in the leather and textile industries.

The procedure involving direct synthesis is the condensation of carbon monoxide with water, and also its reaction with alcohols followed by hydrolysis. The equilibrium is displaced to the product side by bases, for example, NaOH.

$$CO + H_2O \rightarrow HCOOH$$

$$CO + ROH \rightarrow HCOOR$$

The conditions are 8–30 bar at 115–150°C. The conversion of the esters to the acid has to be through the amide to avoid re-esterification of the methanol–formic acid mixture (see Weissermel and Arpe, 1978, pp. 38–39).

Formic acid is also formed in various oxidative degradation reactions, such as oxidation of light naphtha to acetic acid. The first processes are now favored.

E. Acetylene

Acetylene is prepared by two processes which are energy intensive:

1. Coke and lime are heated to 2200°C in an electric furnace to give calcium carbide. Reaction with water gives acetylene:

$$CaO + 3C \longrightarrow CaC_2 + CO$$

$$CaC_2 + 2H_2O \longrightarrow CH{\equiv}CH + Ca(OH)_2$$

2. There are many processes for C_2H_2 manufacture based on the uncatalyzed cracking of hydrocarbons from methane to crude oil. The reaction temperature is >1400°C, the residence time has to be extremely short, and the product gases quenched rapidly:

$$2CH_4 \longrightarrow CH{\equiv}CH + 3H_2$$

4.9 SUMMARY

The aim in this chapter has been to acquaint you with a few examples of the smaller organic molecules called petrochemicals which are one or two steps away from the seven basic petroleum source chemicals. The latter materials are ethylene, propylene, butene, benzene, toluene, xylene, and methane. Processes have been devised for the isolation of these materials in tremendous quantitites from crude petroleum. Emphasis has been placed on a representative selection of manufacturing processes that give an idea of how these enormous quantities of petrochemicals are prepared. Chapter 10 gives an introduction to larger molecules made from these building blocks.

REFERENCES

Andrussow, L. *Angew. Chemie*, **48,** 593 (1955).

Balker, A., W. Capnez, and W. L. Holstein, *Ind. Eng. Chem., Prod. Res. Dev.* **22,** 217 (1983).

Berty, J. M., Ethylene oxide synthesis, in B. Leach, Ed., *Applied Industrial Catalysis*, Vol. 1, Academic, New York, 1983, pp. 202–236.

Cropley, J. B., L. M. Burgess, and R. A. Luke, *CHEMTECH*, **14,** 384 (1984).

Danby, D. E., and C. R. Campbell, in R. Kirk and D. Othmer, Eds., *Encyclopedia of Chemical Technology*, 3rd ed., Wiley, New York, 1978, vol. 1, p. 516.

Dombek, B. D., *J. Chem. Educ.*, **63,** 210 (1986).

Dry, M. E., *CHEMTECH*, **12,** 744–750 (1982).

Eby, R., and T. C. Singleton, Methanol Carbonylation to Acetic Acid, in B. Leach Ed., *Applied Industrial Catalysis*, Vol. 1, Academic, New York, 1983.

Falbe, J., and H. Bahrman, *J. Chem. Educ.*, **61,** 961 (1984).

Farmer, J. B., Inorganic Cyanogen Compounds, in R. Thompson, Ed., *The Modern Inorganic Chemicals Industry*, The Chemical Society, London, 1977, pp. 403–418.

Fisher, W. B., and L. Crescentini, in R. Kirk and D. Othmer, Eds., *Encyclopedia of Chemical Technology*, 3rd ed., vol. 18, Wiley, New York, 1982, p. 430

Forster, D., Rh- and Ir- Catalysed Methanol Carbonylation, in F. Stone and R. West, Eds., *Advances in Organometallic Chemistry*, Vol. 17, Academic, New York, 1979, pp. 255–267.

Forster, D., and T. W. Dekleva, *J. Chem. Educ.*, **63,** 204 (1986).

Francoisse, P. B., and F. C. Thyrion, *Ind. Eng. Chem., Prod. Res. Dev* **22,** 542 (1983).

Frank, M. E., *CHEMTECH* **12,** 358 (1982).

Grasselli, R. K., and J. D. Burrington, *Ind. Eng. Chem., Prod. Res. Dev* **23,** 393–404 (1984).

Grasselli, R. K., *J. Chem. Educ.*, **63,** 216–221 (1986).

Gates, B. C., J. R. Katzer, and G. C. Schuit, *Chemistry of Catalytic Processes*, McGraw-Hill, New York, 1979, p. 139.

Hardy, E. E., in R. Kirk and D. Othmer, Eds., *Encyclopedia of Chemical Technology*, 3rd ed., Vol. 17, Wiley, New York, 1982, pp. 416–428.

Harris, M., and M. Tishler, *Chemistry in the Economy*, American Chemical Society, Washington, D.C., 1973.

Henrici-Olive, G., and S. Olive, *CHEMTECH*, **11,** 746 (1981).

Hillebrand, W. A., *Ind. Eng. Chem. Prod. Res. Dev.*, **23,** 476 (1984).

Innes, R. A., and H. E. Swift, *CHEMTECH*, **11,** 244 (1981).

Inui, T., T. Ishihara, N. Morinaga, G. Takeuchi, H. Matsuda, and Y. Takegami, *Ind. Eng. Chem. Prod. Res. Dev.*, **22,** 26–30 (1983).

Jira, W., W. Blau, and D. Grimm, *Hydrocarbon Processing*, 97 (1976).

Kollar, K. A., *CHEMTECH* **14,** 504–511 (1984).

Krieger, *J.*, *Chem. Eng. News*, Aug. 1, 41 (1983).

Landau, R., G. A. Sullivan, and D. Brown, *CHEMTECH*, **11,** 602–607 (1979).

Marschner, F., and F. W. Moeller, Methanol Synthesis, in B. Leach, Ed., *Applied Industrial Catalysis*, Vol. 2, Academic, New York, 1983, pp. 215–243.

Mol, J. C., and J. Moulijn, *Adv. Catal.*, **24,** 131 (1975).

Mol, J. C., *CHEMTECH*, **13,** 250 (1983).

Morrison, R. T., and R. N. Boyd, *Organic Chemistry*, 4th ed., Allyn and Bacon, Inc. Boston, 1983.

Naworski, J. S., and E. S. Velez, "Oxychlorination of Ethylene", in B. Leach, Ed., *Applied Industrial Catalysis*, Vol. 1, Academic, New York, 1983, pp. 239–273.

Neier, W., and J. Woellner, *CHEMTECH*, **3,** 95 (1973).

Parshall, G. W., *Homogeneous Catalysis*, Wiley-Interscience, New York, 1980.

Prins, R., Reaction Mechanisms in Homogeneous Catalysis, in R. Prins and G. Schuit, Eds. *Chemistry and Chemical Engineering of Catalytic Processes*, Martinus Nijhoff/ Dr. W. Junk, Dordrecht, The Netherlands, 1980, pp. 422–425.

Pruett, R. L., *J. Chem. Educ.*, **63,** 196 (1986).

Putscher, R. E., in R. Kirk and D. Othmer, Eds., *Encyclopedia of Chemical Technology*, 3rd ed., vol. 18, Wiley, New York, 1982, pp. 353, 357.

Rylander, P. N., Catalytic Processes in Organic Conversions, in J. R. Anderson and M. Boudart, Eds., *Catalysis: Science and Technology*, Vol. 4, Springer-Verlag, Berlin, 1983, pp. 1–38.

Sandridge, R. L., and H. B. Staley, in R. Kirk and D. Othmer, eds., *Encyclopedia of Chemical Technology*, 3rd ed., vol. 2, Wiley, New York, 1978, p. 362.

Short, J. N., *CHEMTECH*, **11,** 238 (1981).

Stiles, A. B., Dehydrogenation and Oxidative Dehydrogenation, in B. Leach, Ed., *Applied Industrial Catalysis*, Vol. 2, 1983, p. 162.

Taniguchi, B., and R. Johnson, *CHEMTECH*, **9,** 502 (1979).

Thompson, D. T., "Carbon–carbon bond formation. I. Carbonylation," in R. Pearce and W. R. Patterson, Eds., *Catalysis and Chemical Processes*, Wiley, New York, 1981, pp. 175–179.

Tolman, C. A., *J. Chem. Educ.*, **63,** 199–201 (1986).

van der Baan, H. S., The Acrylonitrile Process, in R. Prins and G. C. Schuit, Eds., *Chemistry and Chemical Engineering of Catalytic Processes,* Martinus Nijhoff/Dr. W. Junk, Dordrecht, The Netherlands, 1980.

van Hooff, J. H., Industrial Catalytic Partial Oxidation Processes, in R. Prins and

G. C. Schuit, Eds. *Chemistry and Chemical Engineering of Catalytic Processes*, Martinus Nijhoff/Dr. W. Junk, Dordrecht, The Netherlands, 1980, pp. 507–521.

Villasden T., and H. Livbjerg, *Cat. Rev. Sci. Eng.*, **17,** 203 (1978).

Weissermel, K., and H.-J. Arpe, *Industrial Organic Chemistry*, Verlag Chemie, Weinheim (1978).

Wing, M., *CHEMTECH*, **10,** 20 (1980).

Wittcoff, H. *J. Chem. Educ.*, **60,** 1044–1047 (1983).

Wittcoff, H. and B. G. Reuben, *Industrial Organic Chemicals in Perspective*, Wiley-Interscience, New York, 1980.

EXERCISES

In answering the following questions, the catalyst information given in Chapter 11 should be kept in mind when you are reviewing the whole course.

1. Describe conversion of petroleum into seven purified chemical substances, using various separation and catalytic conversion processes.

2. Discuss industrial processes using ethylene chemistry. For oxidation processes, discuss the evolution of the current vinyl chloride process, its process details (explaining the process diagram), the catalyst system, and the mechanism of reaction.

3. Discuss industrial processes for vinyl acetate and acetaldehyde as in item 2 but in less detail.

4. Discuss butanol and acetic acid made from acetaldehyde, processes that are now partly superceded by butanol from hydroformylation of propylene, and acetic acid from methanol. Discuss preparation of methanol. Discuss the need to continually search for the cheapest best process.

5. Describe the processes for ethylene oxide, ethylene glycol, styrene (two processes, one a metathesis process), and propionaldehyde.

6. As examples of industrial processes using propylene chemistry, describe (1) acrylonitrile detailed process using ammoxidation, with HCN an important by-product, and (2) process for styrene and propylene oxide in one reactor (Oxirane process). Give several other examples of two-product processes.

7. As an example of an industrial process using C_4 chemistry, describe the process for adiponitrile and its reduction to hexamethylene diamine; also *t*-butanol, butylated phenols, methyl ethyl ketone.

8. Discuss the industrial preparation of the following substances from benzene: adipic acid, caprolactam, aniline (two processes), styrene, and phenol.

9. Tell how the chemistry of toluene is dominated by its being formed from petroleum in too large quantities, and describe the conversions to other useful aromatics. Also describe the preparation of an important monomer, TDI.

10. Discuss the following materials derived from methane: CH_3OH, HCHO, DMT, methylamines, MMA, acetic acid, gasoline by the Fischer–Tropsch synthesis, HCN, cyanuric chloride, chloromethanes, and HCOOH.

5

Conversion of a Laboratory Process to a Pilot Plant and Then to a Plant Procedure

This chapter shows how a process has to be modified when large-scale production is contemplated. So far in this book we have looked at many processes run on large scale, and the reader should be somewhat aware of how different large-scale work is from preparations on a laboratory scale. We now discuss "scale-up," a phrase that refers to conversion of a laboratory process to the manufacturing stage.

5.1 SCALE-UP OF A LABORATORY PROCESS FOR COMMERCIAL PRODUCTION

In order to give the proper perspective on how research and development is carried out, we first discuss the way that an industrial research and development operation is organized. A chemical company needs to have products to sell. These products must be needed by society. The last 50 years have shown that these products are often on the market for a relatively few years. They are then displaced by other products, which either perform better or are less expensive. We now show what a company has to go through to have a flow of products for market.

5.1.1 How a Company Is Organized to Bring Products from the Laboratory to the Commercial Stage

Management must have the means to learn about needs of consumers that can be met by products the company can supply. The author is most familiar with companies in which essentially all the aspects of this procedure are within the corporation. Some companies purchase technology necessary for a new product, (e.g., the licensing of the patent for a product). This chapter will be concerned with the research, development, sales, production, patent, and other branches that get in the picture when new products are being created.

5.1.2 Stage I: Management Defines Project and Approves the Working Plan of Attack

Industrial research chemists do not usually define their own projects, as those working in an academic environment do. The research staff is hired because of its expertise as inorganic or organic chemists, biochemists, and so on. The direction of research takes into account available raw materials and proprietary knowledge of general groups of compounds obtained by the company over many years. Examples of this would be Dow Chemical Company's association with organic and inorganic bromine and chlorine compounds and Merck's association with pharmaceuticals. Both companies have prepared and tested substances within their area of knowledge for many years and have a great many successful compounds on the market. Within the limitations described, the research group endeavors to discover a useful product to fulfill a certain function within the aims of corporate research and development. These objectives are continually modified and refined by published data in technical journals and by input from sales, technical service, and ideas from the research staff. (See pages 1–2 for a general list of product groups included in the areas covered by all branches of the chemical industry.) Substances are prepared by the research people and are examined to see if they are of interest to management. Often many thousands of products are prepared, only a few of which may eventually be of interest. The direction of this synthesis work is restricted only by the background of the scientific staff, the company's patent situation, and the availability of certain raw materials and equipment. The chemist synthesizes products whose structure is considered to have the desired properties and submits the product for screening tests. The compound is carefully purified to ensure that the test results have not been affected by by-products.

5.1.3 Stage 2: Preliminary Cost Calculation

The products that pass preliminary screening are submitted for cost calculation. Two calculations are prepared (see Chapter 6).

Calculation 1. For the first computation, the chemist assumes that a commercial-sized vessel is available (e.g., 2000 L). It is assumed that the lab process is carried out in the filled reactor to give the same yield, using the same solvents and mole ratios of reagents and commercial costs for the raw materials, if available. The only thing changed would be a bigger batch resulting in a larger quantity of material prepared in essentially the same reaction time. This means that, for the same man-hours, more product would be prepared, so the product would be significantly less expensive than the lab preparation. Loading and unloading the reactor will clearly take longer than in the lab and give a slight cost increase.

Calculation 2. For the second calculation, the chemist is asked to make a reasonable estimate, if additional laboratory work where carried out, of how high a yield figure might be reached, whether the time cycle might be shortened, and whether other simplifications might be achieved. The cost calculation is then repeated to see whether the cost would drop to an interesting value if the chemical predictions proved successful. Management then makes a considered judgment as to whether the product would fit into its product line at the optimistic price. If so, it authorizes laboratory process development work.

5.1.4 Stage 3: Laboratory Process Development

This work consists of the following stages:

1. Raw materials are tested in the lab process using samples of material available in commercial quantities. This is because if raw material is sold in commercial quantities only in purity requiring extensive purification before it can be used in this procedure, it will cause the estimate of cost to rise significantly. The same result might occur if the material had to be synthesized.

2. Solvents are to be used that are not severe fire, explosion, or toxicity hazards. Solvents that are dangerous to handle would likely require special safety equipment when used in large quantities. Usually, however, a substitute solvent can be found that has fewer problems.

3. All solvents used in large-scale work have to be recovered for reuse. The recovery process has to yield solvent of usable quality. The solvent should be recovered in at least 85% yield. A loss of 15% is usual in this work. This loss, plus the recovery equipment needed and extra time to be spent, will increase the cost. (For background on distillation set-ups see Chapter 8.)

4. Yield improvement studies must be conducted. A higher yield means lower cost (less kilograms of raw material becoming waste). The process called statistical design is a method whereby a limited number of experiments can be used to identify the significant variables affecting the yield. The studies by Davies (1978) and Hunter (1974) (on the Plackett-Burman procedure) are appropriate in this matter. When these are known, the process can be adjusted chemically in a logical way.

5. Identify all significant by-products chemically. It is necessary to have samples of these materials to add to the product to see what tolerable limits of by-product are acceptable in the finished product in the application tests. The toxicity of the by-products must also be determined. This will prevent situations

where there is a highly toxic by-product in a relatively nontoxic product from going unnoticed. Analysis of the statistical experiments referred to previously for these contaminants shows how the process variables affect the quantity of by-product. For example, a product added to give color stability to plastics should not have any color-forming by-products. The plastic is prepared at high temperature, and since most substances discolor at raised heat, the prepared material would be spoiled.

6. For a particular time cycle and reaction vessel size it is important to prepare as much product as possible. Reactions run in solvent should be studied to have as little solvent as possible, with the vessel as full as possible with reaction mixture, so that the quantity of product in the vessel is as high as possible.

7. Crystallization for purification should be carried out only as a last step. The crystallization process gives a yield of about 90% at best. A 10% yield loss is quite high. Of course, the material in the mother liquor can be purified with additional processing. There are other purification methods that are much more quantitative (e.g., extraction or chromatography).

8. The following factors should be considered in order to have as short a time cycle as possible:

a. Filtration of a slurry has to be fast. This requires well-formed crystals. If a process involving a slow filtration is carried out on a large scale, liquid will stop flowing off when the cake is several inches thick. Studies on how to obtain well-formed crystals must be made. Reduction in the quantity of by-products present is one of many factors that affect the size and quality of crystals.

b. The most common purification method used in large-scale processing is extraction by partition between immiscible liquids. The separation of liquids often requires a long time because of the tendency for emulsification. Laboratory studies to improve the speed of separation should be carried out. Solvents used may have to be changed or temperature adjusted, or a preliminary purification may have to be carried out by another method.

c. To get a clear idea of how the lab process will behave in large scale, the procedure should be carried out using the time cycle estimated for the large-scale preparation. For example, heating up a 4000-L vessel full of reaction mixture may take one-half to three-quarters of an hour, as against five minutes for the corresponding lab run. This extended time may affect the result of the run. Filtering the laboratory batch may be complete in five minutes but may take 1–2 hours in the plant. Will the reaction product be stable under these conditions? If it is not, then process modifications will be necessary.

d. It is necessary to balance the economic need for a short time cycle against the ability to control the increased heat safely when reagents are mixed. An addition rate much slower than that used in the laboratory is often needed so that the cooling water can remove the heat generated in the vessel. As the batch size increases, the cooling has to be more efficient. Equipment change may be necessary. This sort of situation will be discussed in Chapter 8.

e. The final process should be carried out at a temperature 10–20°C higher than the final lab process calls for. This is to see whether the process is close to the point where an uncontrollably fast reaction occurs.

f. Analyses should be developed for showing the rate of product and by-product formation and of disappearance of raw materials. These analyses have to be fast, so that what is happening in the large-scale batch is quickly known to the chemist.

g. All gas, liquid, and solid effluent has to be weighed and analyzed, and a decision on proper disposal must be made. It is often found that additional processing is necessary.

5.1.5 Stage 4: Preparation of Pilot Plant Process

Management continually reviews the data that have accumulated and at a certain time decides that pilot plant studies should be made. Before this decision is made final, there are two more steps.

A written pilot plant process is prepared that includes the large-scale time cycle deemed necessary and the choice of equipment needed.

A second pair of cost calculations is carried out, using the same approach as in Section 5.1.3.

Based on the results of these costs and on a review of the process as well as on safety considerations and availability of raw material and equipment, management makes a decision about whether to approve a pilot plant campaign.

5.1.6 Stage 5: Management Approves Pilot Plant Work

It should be kept in mind that the pilot plant batch is a small, commercial-sized batch and is being examined to see whether there are any obstacles to full-scale production. The process should be the one used for the cost calculation referred to earlier. Pilot plant batches should have many samples taken from the batch at various stages and the process completed in the laboratory. This will clearly show what to do if there are large-scale problems that require process modification. Problems seen only at the pilot stage are:

a. As the vessel size increases, the surface-volume ratio changes significantly. Factors such as reaction rate of a surface-promoted reaction or crystallization rate may vary greatly.

b. Filtrations may be unexpectedly slow. The effect of different filter media is hard to predict. Crystal size and rate of growth are likely to be different when large batches are prepared under different types of agitation conditions.

c. The separation of layers may be slow because of the synthesis of emulsion-forming by-products.

d. Samples of the final product mixture should be taken as the crystals form, and they should crystallize in the laboratory. Experience has shown that changes in conditions such as we have been talking about often result in a different polymorphic form being precipitated. This causes trouble in any product that is sold for use in solid form. For example, if a pharmaceutical active ingredient has been formulated into pills, using a large laboratory batch, and the pilot plant material then is crystallized in another form, it is likely that the new pills will have different characteristics and will not hold together. It may also be that the new active ingredient will have different physiological properties. The rate of solubility in body fluids could be different for the second crystal form.

e. It is often necessary to transfer hot slurries of reaction mixture from one vessel to another. An example of this would be to go from a glass-lined reactor to a stainless steel one when the pH value changes from strongly acidic to strongly basic. It has to be decided at what temperature the transferred mixture should take place in order to prevent the mixture from crystallizing out.

f. Studies to use automated heating and cooling equipment certainly should be made in the pilot plant. Before these devices can be used, the heat behavior should be studied with the manually controlled large-scale batch.

g. The effluent data from the lab should be confirmed for the pilot process. Because of the various changes referred to earlier, extra washes are often necessary. Also a detailed review of the experiences as far as safety goes is essential.

5.1.7 Stage 6: Management Approves Commercial Production

The following is a list of what management will look for before approving commercial production:

a. The pilot plant material from a few batches has to be analyzed and specifications must be agreed on. The figures on the maximum of by-products allowed and the minimum product purity acceptable should be similar to the lab material tested previously but may be slightly different if the data from (b) is satisfactory.

b. The performance of pilot plant material in application tests has to be similar to previously tested lab material.

c. The cost calculation using the pilot plant process rewritten to reflect the large vessel experience should be prepared. It should not be too different from previous values and should present a promising economic picture.

d. A review of the pilot plant process should show it to be safe. A safety data sheet should be assembled for the process. An outline of this document is given in Section 5.1.7.A.

e. Adequate raw material and large-scale equipment has to be available to manufacture the quantity of material to meet expected demand.

A. Safety Data Sheet

The raw materials, intermediates, and products should be tested so that safety data will be available for various emergencies. This information should be on hand where the substances are handled.

Physical Properties. Liquid or solid, boiling point, solubility in water, density and flammability are obvious points a fire crew would need to know.

How to Extinguish the Burning Substances. This involves clear instructions on obvious factors such as, for example, not to extinguish burning toluene with water (fire would spread) and not to use water near substances highly reactive with water (e.g., metal hydrides).

First Aid to Lungs, Skin, and Eyes. Chemical and medical personnel have to work together on this. If a toxic substance comes in contact with the skin, it should be removed by dissolving or suspending it in a liquid that will not cause the substance to be more readily carried through the skin. Skin, eyes, and lungs have to be considered for each chemical. Water is used on the body exterior in most cases.

Chemical Reactivity. One must consider which substances should not be stored in the same area. For example, acids should not be stored with bases, nor oxidizing agents with reducing agents. The author was once called to the scene of a truck accident, where the driver heard a crash in the rear of his rig. He opened up the truck and was overcome by the fumes. Fortunately, this occurred in front of a fire station. Fireman flooded the truck with water. Inspection of the truck showed that is contained a large glass vessel of nitric acid that the driver had placed next to a metal container of sodium cyanide. Because of the hot summer day, the glass container of nitric acid broke. The container of sodium cyanide had not yet corroded through. If the fire department had not responded in a matter of minutes, the heavily built-up area would have been exposed to hydrogen cyanide fumes. Careful planning is needed to avoid problems in handling emergency spills of chemicals during transportation accidents and in emergencies in warehouses. Chemists and fire personnel have to work together in planning adequately for reasonable countermeasures.

Storage of Chemicals. The majority of chemicals have only limited stability. It is necessary to purchase only what will be used up before degradation sets in. This involves arranging for a proper analysis schedule. If material becomes off-grade during storage, a difficult disposal problem can occur, as well as the loss of money.

Ecological Problems. Today the ecological problems of chemical companies are very serious. The treating of processes to reduce pollution will be discussed in Chapter 7, along with the results of air and water pollution.

5.2 SUMMARY

This chapter has listed how a process is changed in necessary ways when large-scale production is contemplated. It has described how the chemist tackles the job, so that he will learn the pitfalls in manufacture. He has been warned to use cost calculations to guide his work. He has been warned about needless hazards, such as using unsafe solvents, and about the need to understand the job of manager of the project who has to see that the final process will be satisfactory to all divisions of the company.

REFERENCES

Davies, O. L., *Design and Analysis of Industrial Experiments*, 2nd ed., with corrections, Longmans, New York, 1978.

Hunter, J. S., *Introduction to Statistics*, Westinghouse Learning., 4th printing, Press Publications, Princeton, NJ, 1974.

EXERCISES

1. What divisions of a company are involved in bringing a new product from research lab successfully to production?
2. What background should a company have in chemical manufacture to consider undertaking the discovery and introduction of a new material to the market?
3. What sort of plans do the research people make in setting out to discover a new product?
4. What is the next step required after research has located this product?

5. What are the activities of the process development group concerning promising products dealing with process, effluent, analysis, and so on?
6. Assuming process development reports an interesting material, what is the next step?
7. What kinds of chemical facts are discovered in the pilot plant that cannot be predicted from lab work?
8. What should be the basis for management to order commercial production?
9. What questions should an investigator ask to decide whether adequate information is on hand at a chemical manufacturing location for the safe handling of certain chemicals?

6

Cost Calculations

This chapter discusses the method used in calculating the cost of carrying out a particular chemical preparation. This is rarely referred to in a typical chemistry course, but it is an important part of the chemist's training.

An academically trained person might feel that it is not important, when the detailed chemistry of a process is being worked out, to be concerned with the cost of the work under way. The identity of the product will be the primary concern. Once that has been proved, the purity will be examined carefully. Very sophisticated methods probably will be used for this. Then the yield is examined critically because most research work on structural or mechanistic work requires a knowledge of how the product was made, and a high yield is important in making this decision. In kinetic studies also, the yield should be high. If the industrial person newly graduated from university were asked to make a cost calculation for his initial assignments, he might find this an uninspiring task. The purpose of this chapter is to give the many reasons why cost calculations are important both to academic people (e.g., in grant or budget preparation) and to industrial personnel at all levels in the corporation.

6.1 BASIC BACKGROUND FOR THE PREPARATION OF COST CALCULATIONS

This chapter deals with calculations for the cost of a batch process. For computations for a continuous process, see Holland (1983). Table 6.1 gives a simplified cost calculation sheet for the preparation of a chemical substance. The result will be in dollars per kilogram (or gram) and will include the *total* cost to the organization for carrying out this preparation. We will now discuss the various components involved in the computation.

6.1.1 Essential Parts of a Cost Calculation

The final figure for the cost of the product is made up of eight component fractions, which are as follows:

1. *Raw Materials*—the fraction of the total cost of preparing the material under investigation resulting from the cost of all raw materials. The calculation

is made as explained below using the quantity of materials needed to prepare the amount of product given in the batch size. If there are materials recovered from the process in reusable quality, they would be listed as a credit on the sheet.

2. *Labor*—the fraction of cost due to the labor (man-hours) spent in making the batch of material. This includes such items as solvent recovery time needing person involvement.

3. *Equipment*—the fraction of cost due to the equipment used for the batch. The vessels used have to be paid for. This is done by considering a portion of the cost of each product made as money set aside as payment for the equipment over a stated number of years (often 10 years).

4. *Storage*—the cost of storing raw materials, intermediates, and finished products. This is a distinct part of the cost, especially of low-priced items. Again, the cost of the buildings, tanks, and so on, must be paid for in a stated number of years and is spread over all batches involved.

5. *Analysis*—extensive analytical control carried out on products, by-products, waste streams, and so on. The use of these facilities has to be paid for in computing product cost.

6. *Utilities*—the energy that must be paid for by adding to the cost the amount used for the batch. Consider the quantity of water, steam, and electricity used to carry out this preparation.

7. *Waste Treatment*—the amount of processing needed on waste streams. The cost of each batch of product manufactured should include the cost of processing the waste.

8. *Overhead*—the main purpose of this discussion of costs is to show the way chemical modifications can be used to *directly* affect the final cost figure in a particular process being studied. There are other costs that the company has to meet. The salaries of administrators, company presidents, have to be paid even though they are not directly involved in the batch preparation. Safety personnel, roads, lights, research expense, taxes and insurance, and sales ex-

penses are other examples of this sort of expense. These costs, which are not directly related to the process, have to be paid from the profits and are included on the cost sheet as overhead. This is usually computed as a percentage of the final subtotal. See Table 6.1.

9. *Capital*—consists of other expenses which are incurred at the start of a project and do not recur. These are for buying land, constructing buildings, and buying equipment. The real estate and construction costs will not be considered further. Equipment costs are explained in Section 6.1.2 (Equipment). The fund of money needed to do business, for example, to cover the purchase of raw material before profits have been realized, is called *working capital*. This is also a cost which will not be discussed further as it is not process related.

6.1.2 Discussion of Cost Components

Table 6.1 is an example of an industrial process cost calculation for a batch process. Each line of the calculation will now be explained.

A. *Cost Calculation Sheet—Batch Process*

This sheet is an extremely simple version of a cost calculation. The purpose is to introduce the concept. The company financial people will need a much more detailed calculation. It is suitable for the calculations made at the lab and perhaps early pilot plant stages. It is very important to understand the descriptions of the various parts of the form.

Product Name. It is critical that the cost calculated be for a particular substance at a stated level of purity, in a certain physical form, and from a dated, written process. The reader should realize that crude benzoic acid will cost less than the analytically pure substance. The product name should read, for example, "Benzoic Acid, crystalline, 95% pure." The cost figure will be for 1 kg of material that contains 0.95 kg pure benzoic acid and 0.05 kg of something else.

Process Date. It is necessary to know the process date to be sure that the calculation refers to the details of that particular procedure. The process date relates the actual conditions existing at that time. A decision has been made on the time cycle that relates to the labor requirements, what equipment to use, and so on. As will become clear, any change in these conditions will affect the cost figure.

TABLE 6.1 Cost Calculation Sheet—Batch Process

Product Name and Description

Process Date	Batch size	M.W. Product		Batches/run
Annual prodn	Molar yield	Base yield		Limiting chemical (M.W.)
Name & descr Raw material	Raw material kg/batch	Raw material cost/kg	Total cost of item/batch	Cost of prod/item/kg
Labor	Hrs/batch	Cost/man hr		
Equipment	Hrs/batch	Cost/hr		
Storage Bulk tank	Kg handled	cost/kg		
Control lab	Hrs/batch	Cost/hr		
Utilities Electricity Water Steam	Quantity	Cost/unit		
Effluent liquid Solid waste	Kg processed	Cost/kg		
Subtotal				
Overhead				
Total cost				

Batch Size. This figure is the quantity of finished product that results from the particular expenditure of time, raw material, and so on, given in the procedure. The kilograms stated are those of the actual material. For example, if the product is 95% benzoic acid, the cost will be the value where each kilogram of product contains 0.95 kg pure benzoic acid and 0.05 kg impurity. This point is being labored so that the reader will not think that the calculated price is that for the contained benzoic acid.

Molecular Weight. The molecular weight used is that of the pure substance. It is needed in the calculations of yield. A problem might arise if, for example, the product were a hydrate. A decision has to be made as to whether the value used is that for the hydrated material or for the anhydrous, and the product description should be consistent with this.

Batches per Run. When a product is being made in batch form, it is usually not necessary to clean out the kettle between every batch if a repeat batch of the same product is to be made in it. The amount of material left clinging to the walls is insignificant, and of the same composition as the incoming reaction mixture. Cleaning a large reactor can be very time-consuming. The number of batches in a run is the number that can be carried out without stopping. A batch requiring a cleaning operation should have a calculation that includes cleaning time. Cleaning may well be required if an impurity tends to precipitate out preferentially on the reactor walls. Also if the crystals tend to layer out on the walls, this will take a rinse perhaps with the recrystallizing solvent at the boil to get it out. This sort of problem will increase the cost.

Annual Production. Annual production tells the cost sheet reader a great deal. When one knows the time cycle for the batch, one can calculate how occupied the vessels will be on this particular product and how much additional production can be obtained from this equipment and manpower if demand increases.

Molar Yield. The chemist will have calculated molar yield since his first course in chemistry. It gives the industrial personnel a quick idea of such things as what quantity of products there will be and the possibility for process improvement.

Base Yield. Base yield (kg actual product ÷ kg raw material) uses the weight of the actual product, often less pure than 100%, for which the cost is being calculated. A base yield is figured for each raw material. This figure gives a quick way of relating raw material needs to a particular production requirement. The figure can be considerably greater than 100%.

Limiting Chemical. Again the term *limiting chemical* has the usual meaning that chemists assign. The sheet reader will know that all other reagents can be reduced in process development work except this one.

Raw Materials. The raw material description is of great importance. As with the product, the grade and purity of the substance to be used should be clear to the reader. The cost of the raw material is almost always quoted on an as-is basis, which means that the analysis of the limiting raw material will be needed for yield calculation and for when a decision is to be made as to whether alternate sources of lower cost can be used. It should be noted that raw material from two alternate sources with the same product analysis could easily not behave the same way because of different by-products. These materials might require changed purification procedures for the product workup and change the cost picture significantly. The cost of the synthesis of an expensive raw material can be made using "intelligent guesses" for the necessary procedure to see if in-house preparation is worth considering.

The total cost of each raw material item for this batch size can be calculated as follows:

Example 6-1. Formula for Calculation of Raw Material Cost/kg Product.

$$\begin{pmatrix}\text{Total cost of item}\\ \text{per batch}\end{pmatrix} = \begin{pmatrix}\text{kg raw}\\ \text{material}\end{pmatrix} \times \begin{pmatrix}\text{raw material}\\ \text{cost/kg}\end{pmatrix}$$

$$\begin{pmatrix}\text{Fraction of cost of}\\ \text{product/kg for}\\ \text{each raw material item}\end{pmatrix} = \frac{\text{(total cost of item/batch)}}{\text{(batch size)}}$$

If there is a recovery of material in the process (essential with solvents), it should be entered as a credit on the sheet. Clearly, the recovered material should be equal in quality to the initial material, or at least reusable. There has to be a charge for the equipment time and labor hours for this recovery step.

Labor. It is necessary to determine the number of people needed to carry out the batch. Most of the time this will be one person, because of the automatic lifting and other devices available. Then it will be necessary to decided how many hours he will have to devote to this task and how much time the equipment can be put on automatic control devices. The cost per man-hour is that at the particular plant location and will include the total cost to the company of employing him.

The cost of the product due to labor is calculated in a similar way to the raw materials.

Example 6-2. Formula for Calculation of Labor Cost/kg Product

$$\begin{pmatrix}\text{Total cost of}\\ \text{labor/batch}\end{pmatrix} = (\text{no. men}) \times (\text{man-hr batch}) \times (\text{cost/man-hr})$$

$$\begin{pmatrix}\text{Fraction of cost of}\\ \text{product/kg for}\\ \text{labor}\end{pmatrix} = \frac{(\text{total cost of labor/batch})}{(\text{batch size})}$$

Equipment. The equipment list is that called for in the process. Chapter 8 discusses this in detail. The factors dealing most directly with costs due to equipment are as follows.

It is important to fill up the kettle as far as possible. This is generally considered to be about 80% of the kettle volume. For a given process it is clear that given a particular vessel, a fuller vessel will mean a larger batch size for essentially the same labor. Referring to the preceding equations, the reader will see that this will reduce the cost.

Each piece of equipment is assigned a utilization cost of a certain number of dollars per hour. This has been calculated as follows: Some of the receipts from the sale of products manufactured in the vessel are used to recover the purchase price of the installed equipment. This money is called part of the cost of the product. A decision is made to charge a fee each time a product is manufactured in the vessel. The vessel is assumed to be in use, for example, for 80% of the time. It is decided to pay for the equipment in a certain time, say, 10 years.

Example 6-3. Formula for Calculation of Charge Price per Hour for the Vessel

$$\begin{pmatrix}\text{Charge price}\\ \text{of equipment/hr}\end{pmatrix} = \frac{(\text{installed price of equipment})}{(\text{no. hr in 80\% of 10 years})}$$

$$\begin{pmatrix}\text{Total batch cost}\\ \text{for equipment}\end{pmatrix} = \begin{pmatrix}\text{no. hr to}\\ \text{make batch}\end{pmatrix} \times \begin{pmatrix}\text{charge price}\\ \text{for equipment/hr}\end{pmatrix}$$

$$\begin{pmatrix}\text{Fraction of cost of}\\ \text{product/kg for}\\ \text{equipment}\end{pmatrix} = \frac{(\text{total batch cost for equipment})}{(\text{batch size})}$$

Storage. Unless there are pipeline supplies, it is necessary to store chemical substances in large quantities at the manufacturing site. This has implications for safety, which are discussed in Chapter 8. We now will describe only the two common methods, bulk and tank. Bulk storage utilizes 55 gallon drums, which can be moved about fairly easily by workmen, especially if they are using forklifts. Tank storage is for thousands of gallons and is used when the production rate becomes large. Here it will become necessary to transfer the chemical by pipe to the location at which it will be used. Tank storage is cheaper per kilogram because there are more kilos among which to spread the cost of the vessel and because there is essentially no labor involved. In the bulk method the container of material has to be brought from a warehouse, perhaps up an elevator, and so on. The empty drum has to be removed. This may be a significant cost factor on a low-priced product. (See Chapter 8 for a further discussion.) The storage cost per hour is calculated in the same way as other equipment. The costs of labor, warehouse, and so on are averaged over all materials and a cost per kilogram based on kilograms handled is reached.

Control Lab. Although the number of hours spent in analysis is small, the personnel are likely to be paid at a high rate. Also, the batch is held up while the analysis is being carried out, which results in the loss of valuable time. Automatic analysis devices at the kettle are more and more common. Cost calculations will show whether such items are justified for the particular product.

Utilities. Electricity, steam, and water are supplied on a metered basis. We will not go further into the calculation now, but with a low-cost product this is significant.

Effluent. The price of treating effluent will continue to increase as environmental concerns mount. This is discussed in Chapter 7. For this discussion, the cost of air pollution control is from the purchase and use of scrubbers and electrostatic precipitators. The solid waste requires the expense of landfills, incinerators, and several other means. The liquid effluent often requires on-site treatment, but it can sometimes be mixed with fuel oil and used as an energy source. The cost of this treatment is averaged over the total production for a plant location. The effluent from each process is measured and a cost of *mixed* waste is calculated per kilo. Again the cost of product per kilogram due to this is calculated as in the preceding cases. The best way of reducing effluent costs is to reduce effluent volume. This is achieved by process improvement studies.

Overhead. As mentioned earlier, some of the money made by the sale of the product has to be put toward paying company officials and other nonprocess related costs. The figure is usually reached by calculating a percentage of the subtotal cost at this stage and adding it on as shown.

6.2 USE OF COST CALCULATIONS TO IMPROVE YIELD

At this point in the chapter, the chemist may say that the accounting approach to costs given here does not bear directly on laboratory work, but is of interest only to those in the marketing divisions of a company. The following paragraphs will show ways in which chemical process development work can be helped by these calculations.

6.2.1 Minimum Molar Excess of a Reagent for Largest Yield Increase

A series of runs can be made where the excess of one reagent is increased and the yield is measured. It is obvious that the measured yield will increase and then level off. It should be clear that the cost will decrease because of the larger yield and increase because of the larger amount of reagent added. The cost calculations for this set of runs will show the smallest excess needed for lowest cost, taking into consideration factors such as the processing of added effluent from the added reagent.

6.2.2 Cost Aspects of Shortening Time Cycle

Laboratory work is carried out to obtain a product, and the procedure is often governed by an 8-hour day. The chemical industry usually runs on a 24-hour day, and it is of interest to have the batch time cycle as short as possible if the yield and quality of the product are not affected. A systematic determination should be made to determine from a series of runs which process has the lowest cost for minimum reaction time.

Optimum Use of Equipment as a Means of Decreasing Time Cycle and Cutting Costs. In the laboratory it is usual for one person to be involved in a synthesis. In such situations the laboratory flask is put away while the reaction is worked up. In contrast, it is essential that industrial reactors be in use as much as possible. It is customary to schedule reactions so that when the reactor for the process is emptied, the next synthesis follows along. See Chapter 8, where the aspirin process is described, to see how a process scheme is set up for high kettle utilization.

6.3 USE OF COST CALCULATIONS TO REDUCE POLLUTION

The approach used in the preceding paragraphs also can be used in reducing the amount of effluent, by using the cost calculations as a measure of where to tackle the process.

Estimate how much it would cost to process effluent to recover more solvent or more product or usable by-product, and calculate the cost reduction from this added material. It should be stated here that reduction in the amount of effluent is obviously important even if no cost reduction takes place, because the nation is running out of disposal sites.

Review the reagents in the synthesis. See how much of each reagent ends up in the final product and whether the eliminated substance has commercial value.

The process should be safe yet economic. In addition to having a satisfactory yield, safety factors are being evaluated. Worker safety involves the control of gases liberated in the workplace and the danger of fire and explosion. There is also the question of pollution outside the plant. Today this question involves the probable purchase of automatic control equipment. The judgment of whether the process can afford capital expenditures of this sort is made by cost calculations.

The question of automation of processes should be evaluated by cost. The question is whether it is safe to have a process involving, for example, a heating cycle where a process is heated to the reaction temperature and then held there for a stated time, controlled automatically.

Cost calculations can be used on current processes to see which of a choice of processes is more appropriate. For example, with the evolution of sulfur dioxide from the manufacture of sulfuric acid, there were two ways to be compared to reduce this problem. The first was to have enough absorbers to take care of the acid gases. The other, which clearly is a more fundamental solution to the problem, was to study the catalytic oxidation of the SO_2 to SO_3. This is because it is necessary to have as high a yield as possible and because sulfur trioxide is more soluble in 99% H_2SO_4 than SO_2. See Section 2.4.2A.

6.3.1 Sample Cost Calculation

In Chapter 8 the description of a process for the manufacture of aspirin is given, showing the large number of kettles needed for even this simple one-step operation. Two cost calculations are given in Table 6.2 for this process at a yield of 95 and 85%. This shows the effect of yield improvement on cost. A 10% decrease in yield gives a 12% increase in cost. In this calculation we added the same amount of mother liquor to the batch in each run. The acetic acid is added to dissolve the salicylic acid. As the yield goes down the quantity of acetic

TABLE 6.2 Cost Calculation Sheet—Batch Process

Product Name and Description
Aspirin (Acetylsalicylic Acid)
(Figures in brackets are for an 85% yield)

Process date	Batch size	M.W. product	Batches/run
7/1/85	818 (732)	180	1200
Annual prodn	Molar yield	Base yield	Limiting chemical. MW
1 million kg	95 (85)	118 (104)	Salicylic acid (138)

Name & descr raw material	Raw material kg/batch	Raw material cost/kg	Total cost of item/batch	Cost of prod/item/kg	
Salicylic acid	661	2.50	1653	2.02	(2.26)
Acetic anhydride	537	1.50	806	0.98	(1.10)
Acetic acid	301	1.00	301	0.37	(0.41)
(acetic acid recovered)	271	1.00	271	((0.33))	(0.37)
Mother liquor	301	0.00	000	0.00	(0.00)
				3.04	(3.40)

Labor	Hrs/batch	Cost man/hr			
2 men/batch each 6 hr	12	20	240	0.29	(0.33)

Equipment	Hrs/batch	Cost/hr			
500-gal reactor	6	20	120	0.15	(0.16)
Crystallizer	6	20	120	0.15	(0.16)
Nutsche tank	6	15	90	0.11	(0.12)
Centrifuge	6	18	108	0.13	(0.15)
Drier	6	18	108	0.13	(0.15)
Recovery still	1	30	30	0.04	(0.04)
			150	.71	(.78)

Storage	kg handled	Cost/kg			
Bulk Tank	1770	0.05	88.5	0.11	(0.12)

TABLE 6.2 *(Continued)*

Product Name and Description
Aspirin (Acetylsalicylic Acid)
(Figures in brackets are for an 85% yield)

Control	Hrs/batch	Cost/hr	Total cost item/batch	Cost of prod/ item/ kg	
Lab	2	30	60	0.07	(0.08)
Utilities	Quantity	Cost/unit			
Electricity	10 units	0.05	0.50	—	—
Water	20 units	0.10	2.00	—	—
Steam	10 units	0.07	0.70	—	—
Effluent	kg processed	Cost/kg			
Liquid	79	0.20	16	0.02	(.02)
Solid Waste	none				
Subtotal				4.24	(4.73)
Overhead (20% of subtotal)				0.85	(0.94)
Total cost				5.09	(5.68)

anhydride probably increases in the mother liquor, but this was not worked into the calculation.

1. The raw material cost/kg and labor and equipment cost/hr are not current figures.
2. The recovery unit is set up to collect the mother liquor from 12 batches in a 4000 *l* vessel. This is distilled to recover acetic acid and acetic anhydride. The time taken to recover the acetic acid from 1 batch is 1 hr.
3. The liquid effluent (30 kg nonrecoverable acetic acid plus 49 kg excess acetic anhydride) will have all the by-products from the reaction. It will have to be disposed of, perhaps by incineration. With proper equipment, the acetic anhydride would be recoverable.

4. The raw materials, labor, and equipment contribute most to cost. The process development people should attempt to adjust the process to decrease these figures.
5. The purity of the aspirin is not addressed here. If a reslurry is needed in a solvent, the product cost will increase.

So even though the batch was in the crystallizer for 16 hours, it is only charged for 6 hours, because the vessels were in use all the time.

6.4 SUMMARY

This chapter introduces the basic idea of cost calculations. The method is simple and will be of use only in the process development stage. It is a powerful tool in pinpointing the costly parts of a process, where attention should be focused. For a more elaborate treatment of this topic, see Clausen (1978).

REFERENCES

Clausen, C. A., and G. Mattson, *Principles of Industrial Chemistry*, Wiley-Interscience, New York, 1978, pp. 316–341.

Holland, F. A., F. A. Watson, and J. K. Wilkinson, *Introduction To Process Economics, 2nd ed.*, Wiley, New York, 1983.

EXERCISE

Calculate the cost of either the process you are studying in Chapter 1 for the term paper or of a process seen on the plant tour. In either case, use assumptions for such figures as yield, cost of raw materials, and so on, if they are not available.

7

The Environmental Impact of a Process

It is essential that everyone in the chemical profession has a basic appreciation, for any process, of how much of the added raw material ends up in the by-products. For example, to make a 1 million kg of phosphoric acid will cause the liberation of 2.93 million kg of calcium sulfate and 88,500 kg of hydrogen fluoride as SiF_4. Let us consider various types of process by-products and discuss the relative environmental impact involved. Fermentation processes will have the dead organisms to take care of. This is a problem with the production of an antibiotic in large quantities. It is presumably nontoxic in the environment because it is made up of naturally occurring substances such as proteins. More serious problems are found with such things as the disposal of the sludge from a municipal waste treatment plant because of the possibility of heavy metal contamination and nonbiodegradable waste in it. The most serious environmental discharge problem is that from organic processes (except from nuclear waste). Organic mother liquors are often difficult to handle, because they consist of a mixture of chemicals that are not biodegradable in natural systems. Some of these by-products are from processes that yield products that are designed to interact strongly with the environment (e.g., pesticides and herbicides). The effluent from these procedures can be highly toxic to living systems.

We now discuss the effect of pollution in general on air, water, and land. The reader should realize that there is an enormous literature on this subject.

7.1 AIR POLLUTION

7.1.1 The Systems Present in Nonpolluted Air

Air is composed of a mixture of gases that exists in a layer about 50 miles deep around the earth. Table 7.1 describes the major components of air when it is nonpolluted. The important fact about this table is that there are small amounts of carbon monoxide, sulfur oxides, nitrogen oxides, ozone, nonmethane hydrocarbons, and methane present in unpolluted air. The first four of these are on

TABLE 7.1 Major Components in Clear Unpolluted Air near Sea Level

Component	Concentration in Unpolluted Air
Nitrogen	78.09%[c]
Oxygen	20.94%[c]
Rare gases[a]	0.932%[c]
Carbon dioxide	345 ppm[b,d]
Methane	1.5 ppm[b,d]
Nonmethane hydrocarbons	0–1 ppb[b,d]
Nitrous oxide	300 ppb[b,d]
Carbon monoxide	0.1–0.2 ppm[b,d]
Ozone	20 ppb[b,c]
Nitrogen oxides as NO_2	0.1 ppb[b,d]
Sulfur dioxide	0.1 ppb[b,d]
Other components in unpolluted air: HCl, Cl_2, CH_3Cl, H_2, NH_3, H_2S	

[a]Rare gases = total of Ar, Ne, He, Kr, Xe.
[b]Gases marked are in unpolluted air but can be increased to a polluting level by human activities.
[c]Adapted from Lodge (1978, p. 106).
[d]Adapted from Stern (1984, pp. 30–31).

the list of gases that are considered pollutants when they are present in quantities greater than those stated in the National Ambient Air Quality Standards. (See Table 7.2.) Each of the items CO, SO_x, NO_x, and O_3 has a significant natural source, as is shown in Table 7.3. The nonmethane hydrocarbons will be shown (Section 7.2.4) to be an essential factor in ozone formation, and the methane will be seen to be important in carbon monoxide formation (see Section 7.2.1). Table 7.4 lists the important natural and anthropogenic quantities of these materials which are evolved.

7.1.2 Polluted Air

We must now define polluted air, taking into consideration that there are so many substances present in natural conditions which would be toxic if found in quantities greater than the values stated in the air quality standards (Table 7.2). We use as the criteria of pollution whether or not there is a measurable effect of the substance on humans, vegetation, important material substances, and/or animals. Pollutants rarely go above 2000 ft. They total about 250–300 million

TABLE 7.2 Selected National Ambient Air Quality Standards[a]

Pollutant	Primary (ppm)	Primary ($\mu g/m^3$)	Secondary (ppm)	Secondary ($\mu g/m^3$)
Particulate				
Annual geometric mean		75		60
Maximum 24-hr concentration		260		150
Sulfur oxides				
Annual arithmetic mean	0.03	80		
Maximum 24-hr concentration	0.14	365		
Maximum 3-hr concentration			0.5	1300
Carbon monoxide				
Maximum 8-hr concentration	9	10000		
Maximum 1-hr concentration	35	40000	Same	
Nitrogen oxides				
Annual arithmetic mean	0.05	100	Same	
Ozone				
1 day exposure—maximum 1-hr average	0.12	235		
Lead				
Maximum arithmetic mean average over 3 months	1.5			

[a]From *Code of Federal Regulations,* Title 40, Section 50, July 1, 1985.

tons worldwide, but if distributed evenly they would amount to about 3 ppm. However, because of slow diffusion from and geologic barriers around pollution sources, serious local pollution can occur.

There are five primary air pollutants and one secondary polluting gas, ozone, which is formed in the atmosphere from some of these materials, along with hydrocarbons.

1. Particulate matter.
2. Sulfur oxides.
3. Carbon monoxide.
4. Nitrogen dioxide.
5. Ozone.
6. Lead.

The combination of nitrogen oxides and the nonmethane hydrocarbons mentioned in Tables 7.3 and 7.4, along with sunlight, gives rise to a gaseous air

TABLE 7.3 Chief Natural and Man-Made Sources of Gaseous and Particutate Pollutants

Pollutant	Pollutant Source	Natural Source
SO_2	Fossil fuel combustion	Volcanoes, reactions of biogenic S emissions
H_2S + organic sulfides	Chemical processes, sewage treatment	Volcanoes, biogenic processes in soil and water
CO	Auto exhaust, general combustion	Oxidation of CH_4, forest fires
NO_x	Combustion	Biogenic processes in soil, lightning
N_2O	Small amounts from combustion	Biogenic processes in soil
CH_4	Combustion, natural gas leakage	Biogenic processes in soil, water
Non-CH_4 hydrocarbons	Combustion	Biogenic processes in soil and vegetation
Particulates	All sorts of industrial activity, especially combustion	Hazes, smokes, and dusts
Lead	Leaded gas, old paint	No major sources

[a] Data on gaseous pollutants from Stern (1984).

pollutant group called photochemical oxidants, of which ozone is an important member. The National Ambient Air Quality Standards (see Table 7.2) lists the maximum quantities of these materials permitted in air. Table 7.1 shows them all, except for lead, to be present in nonpolluted air also, but in less than polluting levels.

To put the air pollution question in perspective, four questions should now be discussed.

1. On a weight basis, what source is responsible for the largest quantity of air pollution?
2. Which of the six major polluting materials is present in largest quantity?

TABLE 7.4 Natural and Man-Made Quantities of Air Pollutants Worldwide[a]

	Quantity Evolved × 10^6 (metric tons)	
Contaminant	Major Pollutant Source	Important Natural Source
Gaseous		
SO_2	212	20
H_2S + Organic sulfides	3	84
CO	700	2100
NO_x	75	180
N_2O	3	340
CH_4	160	1050
Non-CH_4 hydrocarbons	40	2×10^4
Particulates	14.5 (USA)	
Lead[b]		

[a] From Stern (1984) for gases; for particulates.
[b] Not found. Average nationwide Pb concentration = 0.32 $\mu g/m^3$. 1.5 $\mu g/m^3$ = air quality standard not yet overtaken for Pb.

3. Is the concentration of pollutants increasing or decreasing?
4. Which of the six pollutants presents the most serious health hazard?

As everyone who reads the daily newspaper knows, the six major pollutants cause serious local problems in many areas. Table 7.5 gives an estimate of the quantities of these materials evolved from man-made (anthropogenic) sources. (See U.S. Environmental Protection Agency, 1979.) The conclusion from this table is that carbon monoxide is the major polluter and that the internal combustion engine is its major source. This leaves the reader with a misleading picture, however.

7.1.3 Relative Toxicities of the Pollutant Substances

The preceding discussion did not take into consideration the question of toxicity. The National Ambient Air Quality Standards (see Table 7.2) shows the relative harm on a weight basis caused by these materials. All the polluting gases present serious health hazards when present in locally polluted areas, for example, carbon monoxide in urban areas of heavy traffic. These materials have different actions and cause harm in different ways to humans and vegetation. The chemical source and control of these materials is now discussed. For further background on this matter see Stern (1984, 347–362).

TABLE 7.5 Nationwide Primary Air Pollutant Sources and Amounts (1976, USA)[a]

	Weight of Pollutant Produced × 10^6 (tons per year)					
Pollutant Source	CO	NO_x	HC[b]	SO_x	Part[c]	Total
Transportation	94	10.2	12.9	0.9	1.2	119.2
Fuel combustion (stationary)	1.3	13.7	1.6	24.2	5.1	45.9
Industrial processes	8.5	0.8	10.1	5.1	6.7	31.2
Solid waste disposal	3.5	0.2	0.9	0.04	0.6	5.24
Miscellaneous	6.6	0.2	6.3	0	0.8	14.0
Total	113.9	25.1	31.8	30.24	14.5	215.54

[a] Data from U.S. Environmental Protection Agency (1979).
[b] HC, nonmethane hydrocarbons.
[c] Part, total particulates.

7.1.4 Structure of Our Atmosphere

We now consider the structure of the atmosphere. As we go up from the earth's surface, the atmosphere becomes less dense. The troposphere becomes colder and colder. At about 11 km above the earth, the chemical composition changes, and reactions giving off heat occur, with the result that the temperature increases with height. This is the area called the stratosphere. At about 51 km the gas temperature decreases with height. We are away from heat-emitting reactions until we reach 90 km, which is the area called the ionosphere. Here heat-emitting reactions occur, giving stable ionic species.

A brief review will now be given of the major atmospheric polluting chemicals, their origins, their natural sinks, and the ways of reducing them.

7.2 CONTROL OF ATMOSPHERIC POLLUTANTS

Under the Federal Clean Air Act of 1970, as amended in 1977, clean air guidelines are now in effect, worked out by the Environmental Protection Agency (EPA). The polluting gases should be kept below the amounts given in Table 7.2, which gives the summary of the National Air Quality Standards. The states were given a deadline to meet these standards. Table 7.6 shows examples of new source performance standards that set maximums for emissions of specific contaminants from stationary (nonautomative) sources. The figures given show the weight of a given pollutant produced per unit weight of the desired product.

TABLE 7.6 New Source Performance Standards (Selected Examples)[a]

Source	Pollutant	Allowable Emissions
Coal burning steam generators (more than 63×10^6 Kcal/hr.		
	Particulates	0.18 g/10^6 cal heat input
	Sulfur dioxide	2.2 g/10^6 cal heat input
	Nitrogen oxides (as NO_2)	1.26 g/10^6 cal heat input
Nitric acid plants	Nitrogen oxides (as NO_2)	1.5 kg/metric ton of 100% acid produced
Sulfuric acid plants	Sulfur dioxide	2.0 kg/metric ton of 100% acid produced
	Sulfuric acid mist	0.075 kg/metric ton of 100% acid produced

[a]Maximum 2-hr average allowable emissions. Compilation of Air Pollution Emission Factors, 3rd ed., U.S. Environmental Protection Agency Pub. No. A-42, Research Triangle Park, NC, Aug. 1977, Part B.

This number is called an emission factor for the substance. There are factors available for a large number of pollutants. There are questions about the accuracy of some of the figures, but the approach seems quite valid for taking an emission inventory of an area.

7.2.1 Carbon Monoxide

Table 7.7 lists sources of carbon monoxide in the atmosphere. It is clear that man-made emissions are small relative to those from natural sources.

Large quantities of methane evolve from marshes and all decaying vegetable matter. This is the process that occurs in sewage treatment plants when the sludge is anaerobically digested (see Section 3.2.4.B). The CH_4 is oxidized to carbon monoxide by a mechanism utilizing hydroxyl radicals. The latter are produced by photolysis of water, and oxidize CH_4 by the following mechanism:

$$CH_4 + \cdot OH \rightarrow CH_3\cdot + H_2O$$

$$CH_3\cdot + O_2 \rightarrow CH_3O_2\cdot + M$$

$$CH_3O_2\cdot \rightarrow H_2CO\cdot + HCO\cdot$$

$$h\nu + H_2CO\cdot \rightarrow H_2 + CO$$

$$HCO\cdot + O_2 \rightarrow HO_2\cdot + CO$$

TABLE 7.7 Sources of Atmospheric Carbon Monoxide[a]

Source	Quantity ($\times 10^6$) (tonnes/yr)	Percent
Methane oxidation	2173	77.6
Oceans	109	3.9
Other natural causes	255	9.1
Man-caused	263	9.4

[a] Data from *Chem. Eng. News,* July 3, 1972, p. 2; *Science* July 28, 1972, p. 339; and *J. Geophys. Res.*, August 20, 1973, p. 5293. Data adapted to emissions data for CO, Table 7.4.

(M = any inert molecule). (See McConnell et al., 1971.) Anthropogenic (man-made) sources contribute only about 10% of the total carbon monoxide emitted into the atmosphere. There are natural sinks for this poisonous gas, as must be apparent. Soils, particularly those covered by natural vegetation, are very efficient at this. Bacteria and particularly fungi in the soil convert the CO to CO_2. This activity was much less apparent in sterilized soils and more noticeable in tropical soils. This data, from several sources, is reviewed by Stoker and Seager (1976). The hydroxyl radical also converts carbon monoxide to carbon dioxide, as shown in the following equation (Griffin, 1977, p. 22):

$$CO + \cdot OH \rightarrow CO_2 + \cdot H$$
$$H\cdot + O_2 + M \rightarrow HO_2\cdot + M$$
$$HO_2\cdot + NO \rightarrow NO_2 + \cdot OH$$

The areas where there is the greatest CO concentration (e.g., downtown city streets or heavy industry combustion sites) provide no soil sink. Man-made (anthropogenic) CO can build up to levels where health problems occur.

There are three distinct situations which produce CO:

$$\begin{aligned} 2C + O_2 &\rightarrow 2CO \\ 2CO + O_2 &\rightarrow 2CO_2 \end{aligned} \qquad (1)$$

Incomplete combustion of carbon and carbonaceous materials is often seen. This occurs because the first reaction is 10 times faster than the second. The available oxygen is not enough for complete combustion, or because of inadequate mixing is not present locally in sufficient concentration.

$$CO_2 + C \rightarrow 2CO \qquad (2)$$

At high temperature CO_2 can react with carbon-containing materials. This can happen in a blast furnace. The CO formed acts as a reducing agent to convert iron ore into iron.

$$2CO_2 \rightarrow 2CO + O_2 \qquad (3)$$

At high temperatures, CO_2 can dissociate into CO and O_2. If the hot mixture of CO and CO_2 is cooled suddenly, the mixture contains a large quantity of CO, as the equilibrium has not had enough time to shift.

A. CO and Human Health

The effect of CO on human health is shown in Table 7.8. The toxicity is related to the amount of carboxyhemoglobin in the blood. The hemoglobin reacts with the carbon monoxide to block the point where oxygen is attached. Table 7.8A shows how the concentration of carboxyhemoglobin (COHb) in the blood is related to the concentration of CO in the air. The data were listed by Stoker and Seager (1976, p. 21) and surveyed people who smoked, or had such occupations as tunnel traffic police. Table 7.8B shows the physiological effects in humans of various levels of COHb. Figure 7.1 shows how the acute effects of CO depend on the amount of CO in the atmosphere, duration of exposure, and type of physical activity. It is taken from Wolf (1971).

B. Control of Carbon Monoxide Pollution

As was shown on Table 7.2, CO is not as much of a health hazard as the other major pollutants. It is less toxic on a weight basis, does not build up quickly in the blood to a toxic level, and has natural sinks available. When it occurs in a confined area, such as downtown city streets, where there are people working over many hours, a serious health hazard does arise.

Auto transportation was shown to liberate by far the most CO. The control of auto emissions represents a major research and development effort, which will be discussed later.

TABLE 7.8A Blood COHb–Air CO Equilibrium[a]

Ambient Air CO Concentration (ppm)	Equilibrium Concentration Blood COHb (percent)
10	2.1
20	3.7
30	5.3
50	8.5
70	11.3

[a] From Stoker and Seager (1977, p. 21).

TABLE 7.8B Effects in Humans of Various Levels of COHb[a]

Demonstrated Effects	Equilibrium Concentration Blood COHb (percent)
No apparent effect	<1
Possible effect—behavioral performance	1–2
CNS effects, time interval, visual, brightness problems	2–5
Cardiac and lung function changes	>5
Headache, fatigue, drowsiness, coma, respiratory failure, death	10–80

[a]Reprinted with permission from P. J. Wolf, *Environmental Sci. Tech.* **5,** 213. Copyright 1971 American Chemical Society.

7.2.2 Sulfur Dioxide

We turn now to sulfur dioxide, a very toxic acid gas. Sulfur dioxide is discussed in the daily newspapers in connection with the acid rain problem (see Section 7.3.4). It dissolves in water to yield sulfurous acid and sulfuric acid when

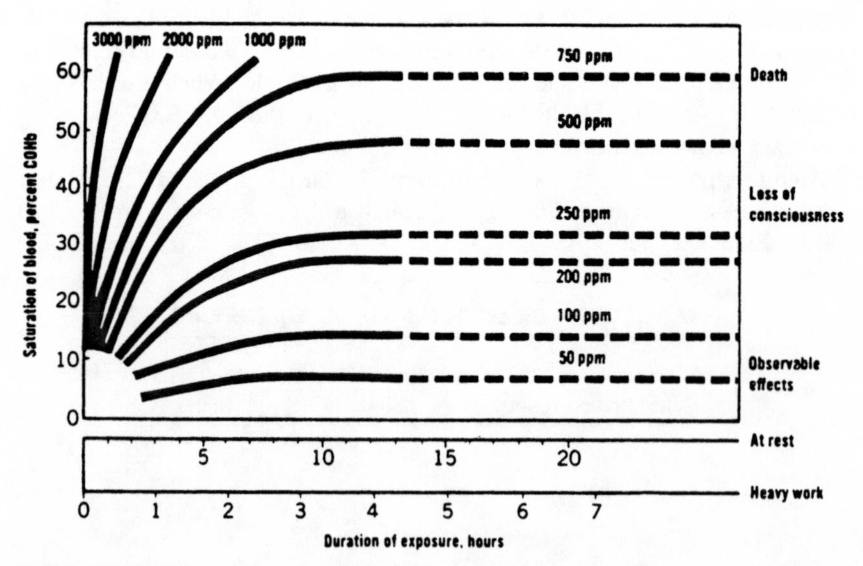

FIGURE 7.1 Causes of acute effects from toxicity of CO. (Reprinted with permission from P. J. Wolf, *Environmental Science and Technology*, 5, 213. Copyright 1971 American Chemical Society.)

oxidation to SO_3 has taken place. A brief discussion of the acid rain problem, which involves several polluting components, is given in Section 7.3.4.

There are two main sources for the sulfur-containing components in the atmosphere. The larger is naturally evolving hydrogen sulfide from decaying organic matter on land and in the oceans. Volcanoes emit SO_2, but 99% of this gas is emitted by man's activities. Most of the H_2S is believed to end up in the stratosphere as SO_2 by oxidation by ozone (Stoker and Seager 1976, pp. 69–71).

$$H_2S + O_3 \rightarrow H_2O + SO_2$$

Ozone is a natural trace element in the atmosphere. It is involved with the nitrogen oxide and hydrocarbon cycles. The reactions of ozone with these components in the atmosphere will be discussed on Section 7.2.4.A.

The second source of SO_2 is from fuel combustion by man. To summarize the origin of airborne SO_2, Table 7.9 is supplied. It is necessary to put in perspective the sources of the anthropogenic SO_2 emitted into the atmosphere in the United States. The yearly emissions are 30 million tons. The primary sources are combustion processes involving coal, oil and natural gas. Coal is the primary culprit. Refining processes involving the latter two materials remove a great deal of the sulfur. Smelting of sulfide ores is another source. Table 7.9 shows the percentage from each source.

A. Chemistry of SO_2 and SO_3 Formation

The following discussion is based on Stoker and Seager (1976, pp. 69–71) and U.S. Environmental Protection Agency (1978, p. 29; 1981, 2-21-2, 2-33). During combustion of sulfur-containing material, both SO_2 and SO_3 are formed. The amount of SO_3 is relatively independent of the amount of oxygen present and varies between 1 and 10% of the total sulfur consumed. This is because the SO_3 is unstable at combustion temperature and because this breakdown occurs more readily with various catalysts. The reaction

$$2SO_2 + O_2 \rightarrow 2SO_3$$

occurs slowly, and thus does not occur to any great extent when the gases are cooled.

The gas mixture from the combustion can be converted to SO_3 by oxidation over vanadium pentoxide. (See the discussion in Chapter 2). In a raindrop formation there is often a tiny dust particle from coal fly-ash. Metallic substances catalyze the SO_3 formation. The stronger the acid in the drop, the more dust that dissolves. This catalytic oxidation proceeds at a much faster rate than without the metal. Other catalysts for the oxidation are called photochemical oxidants, referred to in Section 7.2.5.A.

TABLE 7.9 Origin of Airborne Sulfur Dioxide[a]

Percent Total Atmospheric SO_2	
Oxidation of natural H_2S	55.2%
Fuel combustion	34.3
Industrial processes (e.g., Cu ore smelting)	7.3
Solid waste disposal	0.1
Transportation	1.2
Oxidation of man-made H_2S	1.8

[a] Data from Sources, Abundance and Fate of Gaseous Air Pollutants—Final Report—February 1968, by E. Robinson and R. C. Robbins, Stanford Research Institute Publication PR-6755, p. 18, and also Table 7.5

B. Effect of SO_2 on Plants

Sulfur dioxide either alone or in combination with ozone or NO_2 has a profound effect on growing plants. [For a review of these effects, see Lodge (1978, p. 167–168).] The toxicity data on sulfur dioxide and on the other air pollutants is summarized in Stern (1984, 60–61). The effect of sulfur dioxide on materials is, for example, to attack limestone, the primary building and statuary material. The $CaCO_3$ is converted to $CaSO_4$, which is water soluble and leaches away. [For other data on this source see Stern et al. (1984, Chapter 9)].

C. Control of Sulfur Dioxide Pollution

Twice as much pollution from SO_2 arises from coal burning as from all other activities of man. However, at present we are unable to switch fuels or use an alternate energy source on the scale needed for our long-range energy needs.

It is necessary to tackle the problem of sulfur in coal in two ways, by reducing the sulfur level of coal before burning it or by removing as much as possible of the sulfur oxides from the exit gases before releasing them to the atmosphere.

In order to have low sulfur in utility coal the following should be done:

1. Use coal that analyzes for 0.7% maximum sulfur.
2. Desulfurize high-sulfur coal.
3. Convert the high-sulfur coal to a gas.

We have discussed option 3 previously at some length in Section 2.3.1.A.

7.2.3 How to Manage the Problem of Sulfur in Fuels

A. Methods for the Desulfurization of Coal and Oil before Combustion

There are two approaches to coal desulfurization—physical cleaning and chemical cleaning. Only the former has reached commercial status. The powdered coal is slurried with water, and the pyritic material containing FeS_2 settles out. This fraction in most cases is only a small part of the contained sulfur. There is no commercially feasible method yet available for removal of the organically bound sulfur unless the coal is gasified. If the coal is gasified, then the sulfur can be removed by the Claus process. This was developed for sulfur removal from fuels derived from gasification, liquification, and hydrogenation processes. The sour feed (containing H_2S) and oxygen are passed into a combustion furnace. The resulting gas mixture is sent to a condenser and some of this is recycled to the furnace for temperature control. The rest of the gas is sent to a catalytic converter (containing Fe_2O_3 as catalyst) for partial conversion to H_2S, which then reacts with some of the available SO_2 as shown below to give elemental sulfur.

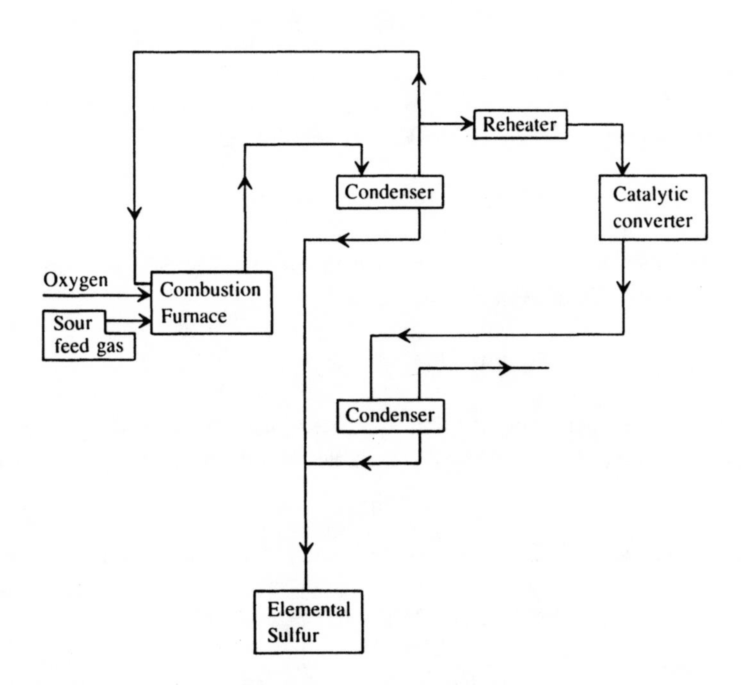

The equations for this process are as follows:

$$H_2S_{(g)} + 1.5O_{2(g)} \rightarrow SO_{2(g)} + H_2O_{(g)}$$
$$SO_{2(g)} + 2H_2S_{(g)} \rightarrow 3S_{(l)} + 2H_2O_{(l)}$$

The mixture containing sulfur is sent to a condenser and the liquid sulfur is collected. The residual gases are sent to another catalytic converter for more conversion of SO_2. More converter–condenser systems are sometimes needed for high-sulfur materials.

Some aspects of this process were mentioned in an update given by Haggin (1986).

B. Removal of Sulfur from Coal Combustion Products

Flue Gas Desulfurization. There are five processes of flue gas desulfurization:

Limestone Scrubbing. Limestone ($CaCO_3$) is mixed with the coal inside the firebox. It reacts as follows:

$$CaCO_3 \rightarrow CaO + CO_2$$

The calcium oxide reacts with the SO_2 and oxygen in the firebox to form calcium sulfate, a substance that can readily be discarded with little environmental effect. Problems are encountered with the $CaSO_4$ coating the walls.

$$2CaO + 2SO_2 + O_2 \rightarrow 2CaSO_4$$

Sodium Sulfite Scrubbing (Wellman–Lord Process). The SO_2 is absorbed in a solution of sodium sulfite in a scrubber.

$$SO_2 + Na_2SO_3 + H_2O \rightarrow 2NaHSO_3$$

The concentrated scrubber solution is heated, and the preceding reaction is reversed, yielding concentrated SO_2, suitable for sulfuric acid or sulfur production.

Citrate Scrubbing. Sodium citrate–citric acid solution absorbs sulfur dioxide.

$$HSO_3^- + H_3Cit^- \rightarrow (HSO_3 \cdot H_3Cit)^{2-}$$

The resulting solution is treated with hydrogen sulfide, and elemental sulfur precipitates. The regenerated citrate solution can be used for further SO_2 removal. H_2S is made from part of the sulfur.

$$(HSO_3 \cdot H_3Cit)^{2-} + H^+ + 2H_2S \rightarrow 3S + H_3Cit^- + 3H_2O$$

Magnesium Oxide Scrubbing. Magnesium oxide scrubbing is the same as the citrate process, in that the dilute SO_2 is absorbed in a slurry of MgO in a scrubber, and the resulting $MgSO_3$ is heated to regenerate MgO and concentrated SO_2 for sulfur or sulfuric acid production.

Catalytic Oxidation. The reader of this book might have thought of this method on his own. We have previously discussed a reaction of SO_2 to prepare sulfuric acid (catalytic oxidation of SO_2 to SO_3 followed by absorption in water). See Section 2.4.2.A. Using concentrated SO_2, a high yield of SO_3 is attained, and the concentrated H_2SO_4 can be prepared. With the same catalyst and the dilute flue gas, 80% sulfuric acid is the highest concentration that has been reached so far.

7.2.4 Nitrogen Oxides

Three oxides of nitrogen are normally found in the atmosphere. These are nitrous oxide (N_2O), nitric oxide (NO), and nitrogen dioxide (NO_2). Natural and anthropogenic (caused by man) sources are known for all three. The origin of these gases is as follows (see Tables 7.3, 7.4):

N_2O—essentially all from bacterial decomposition of nitrogen-containing compounds.

NO—bacterial action gives 80% of the atmospheric NO. The rest is from combustion processes.

NO_2—from oxidation of NO.

Since nitrogen oxides are emitted worldwide by natural processes, one might conclude that they do not present a hazard. However, the situation is similar to carbon monoxide, where serious health hazards can arise in local areas.

A. Nitrogen Oxides as Polluters

In discussing pollution from nitrogen oxide, the term NO_x is often used. This formula is meant to include only NO plus NO_2. This is because:

1. NO and NO_2 are toxic but N_2O is not.

2. NO and NO_2 participate in atmospheric photochemical reactions but N_2O does not.
3. NO and NO_2 are in our air from man's activities. The atmospheric concentration of N_2O from man's activities is insignificant.

Essentially all NO_x from man's activities comes from two sources—fuel combustion from stationary and mobile sources. Other industrial activities emit less than 5%.

B. Chemistry of Nitrogen Oxide Formation

Nitrogen and oxygen are present in the air around us and essentially no reaction occurs. At combustion temperatures, however, these gases do react significantly, as shown in Table 7.10. The equations are as follows:

$$N_2 + O_2 \rightarrow 2NO$$

$$2NO + O_2 \rightarrow 2NO_2$$

The residence time in the combustion zone is important. Note that the temperature has to be in the vicinity of 1200° before the equilibrium shifts to favor NO formation. The exit gases are rapidly cooled to a temperature where the reaction of NO and O_2 is very slow. Only 10% of the nitrogen oxides emitted are present as NO_2. The reaction of NO with oxygen does not take place under combustion conditions. NO_2 decomposes over 600°C to give NO and O_2.

Table 7.10 Theoretical Equilibrium Concentration of NO and NO_2 in Air (50% Relative Humidity) at Various Temperatures[a]

	Concentration ($\mu g/m^3$)	
Temperature (K)	NO	NO_2
298	3.29×10^{-10}	3.53×10^{-4}
500	8.18×10^{-4}	7.26×10^{-2}
1000	43	3.38
1500	1620	12.35
2000	9946.25	23.88

[a] U.S. Environmental Protection Agency Air Quality Criteria for Oxides of Nitrogen, Environmental Criteria and Assessment Office, Office of Research and Development, Research Triangle Park, North Carolina, 1982, p. 3–8.

TABLE 7.11 Oxidation Rate of NO in Air at 20°C in 20% Oxygen

NO Concentration (ppm)	Oxidation Time to Complete		
	25% of Reaction	50% of Reaction	90% of Reaction
10,000	8.4 sec	24 sec	3.6 min
1,000	1.4 min	4 min	36 min
100	14.0 min	40 min	6.0 hr
10	2.3 hr	7 hr	63 hr
1	24 hr	72 hr	648 hr

[a]Reprinted with permission from *Matheson Gas Data Book*, 5th ed., W. Braker and A. L. Mossman, Eds. Copyright 1971 by Matheson Gas Products, E. Rutherford, NJ.

Another experimental result bearing on the ratio of the nitrogen oxides in the atmosphere is summarized in Table 7.11. We see there the dependence of the reaction rate on NO concentration. As the exit gases are diluted, the rate falls off dramatically, and so very little of the nitric oxide emitted would be converted to nitrogen dioxide. (See Braker and Mossman, 1971.)

NO_2 Photolytic Cycle. Nitrogen dioxide is a strong absorber of ultraviolet light from the sun. This interaction is responsible for many photochemical reactions which will be discussed in Section 7.2.4.A. The biological effects of the photochemical products will be discussed.

C. Control of NO_x Pollution

There are two ways to control the quantity of nitrogen oxides emitted from combustion processes:

1. Modify the combustion process to have less NO_x formed.
2. Remove the NO_x from the combustion gases.

Both approaches appear to have some success. Reference to Table 7.10 shows that reduction of the combustion temperature will result in less formation of oxides. This has been achieved by diluting the air–fuel mixture with an inert noncombustible gas. The technique is called exhaust gas recirculation (EGR). The volume of the gas used consists of 15–22% of the air–fuel mixture and is made up of the exit gases from the engine circulated through the EGR valve. The peak combustion temperature is lowered and the oxygen content is decreased in the combustion mixture. To achieve the same effect in a stationary modern electrical power plant, a two-stage combustion is used. In the first stage, the fuel is burnt at a high temperature, with only 90% of the required oxygen for complete combustion. The formation of NO is limited by the oxygen deficiency. In the second stage the combustion is completed at a lower temperature. This again limits the NO formation.

The second approach involves automobile catalytic converters. These will be discussed in Section 11.1.2A.

7.2.5 Hydrocarbons and Photochemical Oxidants

The previous three air pollutants discussed are materials that are toxic substances even when present in small quantities. Hydrocarbons are not especially toxic themselves, but it will be shown that their presence in the atmosphere is the reason for the formation of a series of highly toxic chemicals called photochemical oxidants.

A large number of hydrocarbons get into the atmosphere as gases or in some cases suspended material. They are usually C_{12} or lower in carbon content. (We immediately set aside methane and terpenes from trees, which are not of concern.) It will be recalled that methane can be converted to carbon monoxide by another substance in the atmosphere called hydroxyl radical (·OH), which we have already discussed. Terpenes, although reactive as photochemical substrates, do not need to be considered, because their background concentration in nonurban air is only 0.1 ppm (somewhat higher in deep forests).

The sources of the troublesome hydrocarbons are given in the Table 7.5. It can be seen that cars are the most serious polluters. It must be also be pointed out that industrial process losses, including many petrochemical processes and those involving syngas, are the cause of some of the problem.

A. Photochemical Oxidants—Background Data

Photochemical oxidants are atmospheric substances that are produced by photochemical processes that will oxidize materials not readily oxidized by gaseous oxygen. Ozone and the family of substances called peroxyacyl nitrates are the most familiar of these materials. The most familiar member of the series is peroxyacetyl nitrate. It is called PAN.

The greatest effect of these substances in a biological sense is on plants. Various crops show a dramatic reduction in yield with very small concentrations in the air. Five hours exposure to a PAN level of 0.01 ppm in the atmosphere will injure the most sensitive plants, especially forage, salad crops, and coniferous trees.

Ozone in the 1–3 ppm range has a dramatic effect on humans—extreme fatigue and lack of coordination in sensitive subjects or those who are physically active. Animals show pulmonary damage at this level. A clear example of the human toxicity was found in the first flights of the supersonic aircraft, the

Concorde. The flight attendants complained of fatigue, lack of coordination, and so on. It was found that these symptoms were due to small amounts of ozone that had penetrated the cabin. (It will be recalled that this plane flies in the stratosphere, which has a significant ozone content.) The attendants, walking around, were more subject to these toxic symptoms. The passengers who were sitting down were not bothered. The problem was solved by eliminating ozone by filtering the air through absorbents.

B. Chemistry of Photochemical Oxidants

Many cities today are bothered by a photochemically generated pollution haze. It was first chemically studied and identified for the Los Angeles County area and given the name smog. It is caused by the action of sunlight on mixtures of hydrocarbons and nitrogen oxides. In Los Angeles, dispersion of the pollutants is inhibited by mountains and by the meteorological inversion often seen in the area.

There is a small amount of nitric oxide and nitrogen dioxide present in the atmosphere. Some of this is presumed to originate from N_2O from bacterial action on nitrogen compounds in the soil. The NO and NO_2 are formed from N_2O in the stratosphere, and some comes down to the troposphere.

Nitrogen dioxide is a strong absorber of ultraviolet light from the sun. The following reactions are considered to occur under normal circumstances when no hydrocarbons are present (M is an inert body) (Griffin, 1977, p. 18):

$$NO_2 + \text{light} \ (<430 \text{ nm}) \rightarrow NO + O^* \quad (1)$$

$$O^* + O_2 + M \rightarrow O_3 + M \quad (2)$$

$$NO + O_3 \rightarrow NO_2 + O_2 \quad (3)$$

(O^* is an oxygen atom). The concentrations of NO, NO_2, and O_3 have a constant value.

$$[O_3] = \frac{k_1[NO_2]}{k_2[NO]} \quad (4)$$

$$= 0.021 \text{ ppm} \frac{[NO_2]}{[NO]} \quad (5)$$

where k_1 is the rate constant for reaction (1) and k_2 the rate constant for reaction (3). As long as the ratio $[NO_2]/[NO] < 1$, $[O_3]$ is very low. The situation changes when hydrocarbons are present. We see that the NO_2 concentration builds up significantly relative to that of NO. The ozone concentration increases rapidly also. Equations (1)–(3) cannot explain this.

It is known that the O_3 and O* from Eqs. (1)–(3) will attack various hydrocarbons. However, it is experimentally observed that the rate of loss of hydrocarbons is much greater than would be explained by attack of ozone or oxygen atoms.

The following two proposals (Griffin, 1977, pp. 22–24) are among many that will bring about the oxidation of NO to NO_2 and explain the rapid buildup of ozone.

1. Using CO

$$HO\cdot + CO \rightarrow H\cdot + CO_2 \tag{6}$$

$$H\cdot + O_2 + M \rightarrow HO_2\cdot + M \tag{7}$$

$$HO_2\cdot + NO \rightarrow HO\cdot + NO_2 \tag{8}$$

This sequence, a catalytic cyclic chain reaction, represents an important sink for naturally evolving carbon monoxide. (See Section 7.2.1.A.) The presence of CO has set up a chain reaction that rapidly converts NO to NO_2.

2. Reaction of olefins with HO· (also O_3, $HO_2\cdot$)

$$HO\cdot + C_3H_6 \rightarrow \cdot CH_2CH{=}CH_2 + H_2O \tag{9}$$

$$HO\cdot + C_3H_6 \rightarrow CH_3\dot{C}HCH_2OH \tag{10}$$

Calling the olefin radicals R·

$$R\cdot + O_2 \rightarrow RO_2\cdot \tag{11}$$

$$RO_2\cdot + NO \rightarrow RO\cdot + NO_2 \tag{12}$$

The RO· oxidizes the NO to NO_2. Very little is left for the ozone to react with and so the concentration builds up.

The possible fate of the radical RO· is as follows:

$$RO\cdot + HO\cdot \rightarrow RO_2\cdot + H\cdot \tag{13}$$

$$H\cdot + O_2 \rightarrow HO_2\cdot \tag{14}$$

[This radical is then used in Eq. (8).] Again a chain reaction has been set up to rapidly convert NO to NO_2, using reactions (8)–(14).

The preceding chain reactions can be stopped by the following two types of reactions:

$$R\cdot + R\cdot \rightarrow RR$$

$$R\cdot + NO_2 \rightarrow RNO_2$$

The general formula for the PAN-type products is as follows:

$$R\underset{\displaystyle \overset{\|}{O}}{C}OONO_2$$

A possible reaction sequence for the formation is

$$RCH{=}CHR + O_3 \rightarrow RCHO + \text{various radicals}$$

$$RCHO + HO\cdot \rightarrow RCO\cdot + H_2O$$

$$RCO\cdot + O_2 \rightarrow RC(=O){-}OO\cdot$$

$$RC(=O){-}OO\cdot + NO_2 \rightarrow RC(=O){-}OONO_2$$

The initial source of free radicals is reactions of the type

$$HONO + \text{sunlight} \rightarrow HO\cdot + NO$$

$$CH_2O + \text{sunlight} \rightarrow CHO\cdot + H\cdot$$

$$CH_3CHO + \text{sunlight} \rightarrow CH_3\cdot + CHO\cdot$$

The aldehydic products arise from the ozonization of olefins. Aldehydes are well known in urban atmospheres.

In summary, we have oxidizable pollutants (e.g., hydrocarbons, aldehydes, CO) that cause the formation of free radicals that react with oxygen, giving such radicals as $RO_2\cdot$, and $HO_2\cdot$. These "pump" NO to NO_2 and become degraded to other compounds, some of which are still active because they are free radicals. The amount of conversion of NO to NO_2 and thus the amount of photochemical oxidant formed depend on the reactivity and concentration of oxidizable pollutants. The early smog research focused on the most reactive of the hydrocarbons, which are preferentially consumed. The radical reactions shown with NO leave ozone unconverted, and so its concentration increases.

In summary, we have shown that $[O_3]$ is related to the intensity of the sun and to $[NO_2]/[NO]$. We have shown ways to increase the latter ratio, and so have explained the ozone concentration buildup. Hydrocarbons, aldehydes, ketones, chlorinated hydrocarbons, and carbon monoxide all give peroxy radicals which will react with NO to increase the $[NO_2]/[NO]$, and consequently the $[O_3]$. These materials are typical industrial chemical by-products. This recent work is given to show that no single pollutant can be blamed as the major cause for the buildup of ozone.

7.2.6 Particulate Matter

The four major air pollutants that have been discussed so far have all been gases. Small solid particles and liquid droplets make up a fifth group. Collectively, these are called "particulate matter," and under certain conditions they become a serious pollution problem.

The terms *aerosol* and *particulate* are often used interchangeably. Mist, smoke, fume, and dust are types of particulate matter. [One can also have viable particulates (e.g., bacteria, fungi, molds, and spores).]

At this time, considering the diverse makeup of these materials, it is necessary to consider the various ways the toxicity is shown. This will show why they are considered as a group.

Many particulates penetrate the lungs more effectively than gaseous pollutants. The material goes into the alveoli of the lungs and is retained there. Some particulates enhance the effects of other pollutants in a synergistic manner, some originate from gaseous pollutants, and almost all reduce visibility.

A. Sources of Particulate Matter

We shall discuss two types of anthropogenic (man-made) particulate matter: primary particles and secondary particles. Primary particles are emitted as particles from the source. Such material as soot, fly ash, and dust are examples of this type. Secondary particles are formed in the atmosphere, from gas to particle conversions. Examples of these are solid or liquid particles that are formed when hot gaseous emissions from a utility stack are cooled and the haze that forms on cooling an organic vapor. Sulfate salts from SO_x, nitrate salts from NO_x, and ammonium salts are further examples of secondary particle emissions.

B. Health Effects of Particulates

The National Ambient Air Quality standards give standards based on weight of particles in a stated volume of contained air (75 $\mu g/m^3$, considering human health). The physical and chemical properties of the material are of critical importance. Only those particles smaller than 15 μm (mouth breathing) or 10 μm (nasal breathing) reach the lower respiratory tract. The most serious aspect of this type of pollution is with the particles smaller than 2 or 3 μm. These penetrate to, and probably will be retained by, the alveoli, the deepest part of the lungs, where gases interchange with the circulatory system. Clearly, the shape as well as the carcinogenicity and other toxic properties make this pollution a most serious problem. The Clean Air Act of 1977 was amended to take into account the size of the particles, as well as their chemical composition. Devices that can count particles are being considered as a way of controlling this problem.

C. *Controlling Particulate Emissions*

Reduction of SO_x and NO_x will clearly reduce the secondary aerosols based on these gases. The primary particulates have to be physically removed. There are two methods for this. One uses a filtration device called a baghouse, and the other is a piece of equipment called an electrostatic precipitator. (For a discussion of these devices, see B. B. Crocker et al. (1978)).

7.3 POLLUTION OF THE SOIL: The Systems Present in Nonpolluted Soil

The soil covering the earth is the basis for nearly all the life forms on the planet areas not covered by water. It consists of mineral solids, organic matter, water, and air. The proportions are usually in the area of one-half minerals and solids to one half water and air. Clay minerals and soil provide areas where the above four components interface with each other, and so are excellent for all sorts of life forms.

7.3.1 Weathering

The nature of soil and the physical and chemical changes still occurring in it are described briefly so you will see what should be present naturally and how pollution can change the composition and structure of soil. We will now describe primary weathering. Over eons of time, rocks and minerals that were deep inside the earth and were at equilibrium with their environment came to the surface and adjusted to the new environment of lower temperature and pressure. In this case the minerals that crystallized first from igneous rock magma are preferentially removed from the rock by water, oxygen, and carbon dioxide, an acidic solution. Soils that are at an early weathering stage are the soils of the desert region. The wheat and the corn belt area is considered to be at an intermediate stage. Tropical areas are characterized by low fertility and low plant nutrients.

In secondary weathering, the dissolved ions precipitate out to form silicate and oxide clays. The precipitate is formed in small crystals, colloidal crystals uniquely suitable for soil chemical reactions.

In summary, weathering means the depletion of alkaline and alkaline earth metals, as shown in the following equation:

$$\underset{\substack{\text{Orthoclase} \\ \text{(a mineral)}}}{KAlSi_3O_8} + H^+ \rightarrow \underset{\text{(a clay)}}{HAlSi_3O_8} + K^+$$

The cation is leached away or is taken up by roots.

The structure of clay particles, called micelles, is a three-dimensional array of SiO_4 tetrahedra, all of which share three oxygens with their neighbors. Above this is a layer of aluminum ions. Figure 7.2 shows an example.

7.3.2 The Structure of Soil

Water is associated with the surface cations by hydration and by hydrogen bonding to the oxide ions in the surface. The absorbed water is retained by soils dried at ordinary temperature and is not available to plants. Once the primary weathering has given the colloidal clay, with release of nutrient ions, plants and other organisms can live.

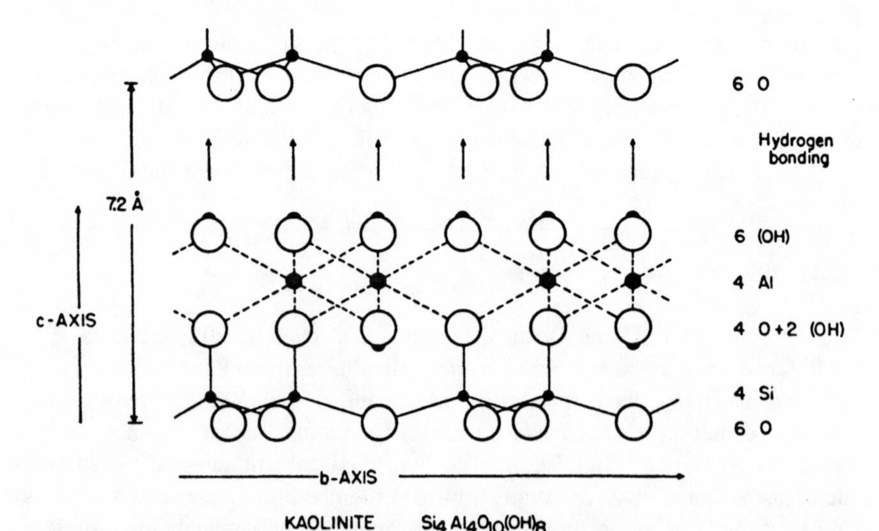

FIGURE 7.2 Structure of silicate clay. (Reprinted with permission from F. E. Bear, Ed., *Chemistry of the Soil* (ACS Monograph Series No. 160). Copyright 1964 American Chemical Society, p. 107.)

The photosynthetic plants as they grow make organic matter, cellulose, proteins, fats, and waxes. When they die they are decomposed by soil organisms. The cellulose and hemicellulose are rapidly decomposed, leaving lignin combined with protein. The lignin is composed of various aralkyl alcohols. The lignin–protein mixture is called humus. It is inert and has a nitrogen content of 3–6%, as peptide linkages. It is insoluble in water, is soluble in acid and base, and (most important) absorbs enormous quantities of water. It also helps to make a healthy soil by providing aeration.

The nutrients for growing plants are obtained as follows: Carbon, hydrogen, and oxygen come from the air, water, or soil. The rest of the nutrients are from the soil. Certain soil ions are absorbed and others are excluded. There is free energy needed to transfer nutrient ions from soil solution to plant fluid. This is supplied by oxidative metabolic reactions in roots. If oxygen is kept from the root area, the uptake of the nutrients can be halted.

A. Ion Exchange in Soil

The macronutrients most likely to be in short supply are those where the form used by the plant is very soluble (e.g., nitrate), those where the plant uses a relatively insoluble form (e.g., phosphate), and those whose cations are less tightly bound to the micelle (e.g., K^+). Nitrogen fertilizers such as ammonium salts or urea are not soluble until the plant oxidizes them to nitrate. In Section 2.1.2 is shown the conversion of the natural phosphate rock to the more soluble superphosphate.

Nutrients will not be available to plants unless they are in a chemical form for which carriers exist. Carriers are molecules that facilitate transfer of the nutrient across the cell membrane. Let us now discuss how nutrients are stored in the soil. They obviously cannot be stored as insoluble compounds, nor as soluble compounds, which would leach away. They are usually held on ion exchange sites on silicate clays and on humus. These are negatively charged micelles and have varying degrees of attraction for cations in the soil. Figure 7.3 (Foth, 1984) tells how ions are held by a micelle, a colloidal-sized clay and humus particle. These are usually negatively charged, and have taken up very large quantities of cations. When fertilizer is added (e.g., KCl), some of the cations are liberated and are available for plants. Figure 7.3 shows how this occurs.

There are other equilibria in soils.

B. Correct Soil pH for Utilization of Nutrients

Certain cations when desorbed from the soil increase pH:

$$\text{Micelle)Ca} + 2H_2O \rightarrow \text{micelle} \begin{matrix} (H \\ (H \end{matrix} + Ca^{2+} + 2OH^-$$

$$\begin{array}{c} 40\ Ca^{+2} \quad 10\ K^{+} \\ \boxed{\quad \text{micelle} \quad} \\ 10\ H^{+} \quad 20\ Mg^{+2} \end{array} + 40\ KCl \rightleftharpoons \begin{array}{c} 36\ Ca^{+2} \quad 32\ K^{+} \\ \boxed{\quad \text{micelle} \quad} \\ 2\ H^{+} \quad 17\ Mg^{+2} \end{array} + \left\{ \begin{array}{l} 4\ CaCl_2 \\ 3\ MgCl_2 \\ 18\ KCl \\ 8\ HCL \end{array} \right.$$

FIGURE 7.3 Equation showing liberation of cations from a clay micelle. (Reprinted with permission from H. D. Foth, *Fundamentals of Soil Science,* 7th Ed., Wiley, New York, 1984, pp. 192.)

The percentage base saturation tells how much of the cation exchange sites are filled by cations. Certain other cations lower the pH when they are released. For example, when aluminum is desorbed by acid (e.g., acid rain), the Al then occupies the cation surface. Some dissolves in the soil solution and hydrolyzes as follows:

$$Al^{+3} + 2H_2O \rightarrow Al(OH)_2^{+1} + 2H^{+}$$

Percent hydrogen saturation tells the amount of the soil covered with this type of ion. The type of cations absorbed on the surface determines the pH of the soil. Variations in soil pH will clearly govern the uptake of nutrients by plants.

C. *Soil Management*

The solid earth has only a limited capacity to take care of various sorts of waste without damaging its ability to grow plants. Removal of harmful substances is brought about by soil bacteria and fungi and by the breakdown of materials when they have been absorbed on the catalytic clay surfaces. It is important to save the soil structure, that is, its ability to hold water and permit air to get to roots. Nutrients can be added to replace those lost, but without the soil structure, erosion will be a serious problem. This is achieved by adding back to the soil a great deal of the crop residue to keep up the level of organic material. In summary, one needs oxygen, water, colloidal surfaces, and microorganisms to grow plants and also to detoxify wastes in soil. For these agents to do their job, it is necessary to have the all-important soil structure.

7.3.3 Adjustment of Process to Reduce Environmental Damage

Procedural changes such as those discussed in Chapter 6 to achieve cost reduction will bring about less pollution. When excess reagent and solvents are used, they must be recovered, not only for reasons of cost but also because the material would damage the environment. The following examples show what is important.

A. Soil Organisms and Environmental Quality

You should consider what effect the compounds being studied will have in soils.

Pesticide Degradation In Soils. Organisms have evolved that can decompose all compounds formed directly or indirectly from natural photosynthesis processes. Pesticides ideally should exert their useful function and then disappear from the ecosystem. It is interesting to note that 2,4-D is readily decomposed by soil bacteria, but 2,4,5-T is very resistant. This is recorded by Foth (1984). It is very important to measure the ease of biodegradation of all agricultural chemicals.

Soil as a Living Filter for Sewage Effluent Disposal. Effluent from sewage treatment plants contains increased nutrients. Plant growth in the receiving water temporarily increases to an artificial amount, but the oxygen will decrease in a short while when the plants die. An experiment has shown that soil will act as a living filter, degrading detergents, and holding nutrients. The filtered water is then used to recharge the aquifer. See Foth (1984).

Oil and Natural Gas Contamination of Soil. Oil spills create an anaerobic soil. The vegetation is killed by lack of oxygen. Soil aggregates are broken down. There are various measures to attack this problem. Cultivation causes aeration and also permits the attack of bacteria on the oil. There are over 100 species of bacteria that will attack petroleum.

7.3.4 The Acid Rain Problem

A consensus has been reached in Europe and North America that acid rain presents a distinct environmental hazard. We will now summarize how industrial chemical processes and other major sources of air pollution are causes of this serious problem.

Normal rain and snow have a pH value of 5.6, due to the presence of dissolved CO_2, which gives carbonic acid. We will now discuss the widespread environmental acidification that occurs from the oxidation of airborne sulfur and nitrogen compounds to form the strong acids H_2SO_4 and HNO_3. The pH has been found as low as 3 to 4, especially in New England and the Canadian maritime areas and in mountainous regions of North Carolina.

The combustion of sulfur-containing fuels, chiefly coal but also oil, gives sulfur dioxide. Only 1–10% of the sulfur is converted to SO_3. There are two major systems present in polluted atmospheres which increase the amount of SO_3 formed. These are metals in ash from the combustion of coal and photochemical oxidants (see Sections 7.2.2.A and 7.2.4.A). The formation of H_2SO_4 occurs by the following reactions of SO_2:

1. In the gas phase with pollutants such as ozone.
2. In or on particles in the atmosphere.
3. In the aqueous phase in fogs, clouds, and rains.

The acidification of ecosystems occurs through the direct precipitation of acids already in the atmosphere or from the precipitation of acid precursors (SO_2, NO_2, H_2O_2, photochemical oxidants, and O_3, which may subsequently react to form acids). A recent review on this complex topic is Gertler et al. (1984).

The removal of sources of sulfur pollution will clearly be a major factor in reducing acid rain. Sections 7.2.2.C and D discuss procedures for removing sulfur both before and after coal and oil combustion. Removal processes for SO_2 are also needed in treating the off-gases from smelting of metal ores. SO_2 is a major local pollutant where this is carried out. In some cases, the SO_2 from smelting is converted to sulfuric acid for fertilizer purposes (see Sections 2.4.1 and 2.4.2). The gasification of coal permits the more efficient removal of sulfur because the sulfur exists mainly as H_2S in the fuel. This can be converted to elemental sulfur by the Claus process (see Section 7.2.2.C). This process is also used in sulfur recovery from petroleum and natural gas.

The source of the nitrogen oxides in the atmosphere is the combination of nitrogen and oxygen at the high temperature that exists in combustion engines, both stationary and those used in transport. This problem has been tackled by modifying the combustion process to reduce NO_x formation (see Sections 7.2.3.B and C) and by treating the combustion product gases to change the nitrogen oxides to nitrogen (Section 11.1.2.A). The NO_2 concentration can build up to dangerous levels from the presence of hydrocarbons and other products from transportation and petrochemical processes. The chemistry is discussed in Section 7.2.4.

The conversion of NO_2 to HNO_3 is discussed in Section 2.5.

The acid rain appears to have widespread effects on trees, and also on fish life in lakes. The dissolution of metals, particularly aluminum, from clays by the acidified rain may be a key factor in this (Section 7.3.2.B). A major international effort was announced in March 1986 by the governments of Canada and the United States to tackle this problem.

In order for the reader to realize that this is a complex issue, the author recommends that the article by Innes (1984) be studied. This is a survey of what the effects of acid rain really are and what economically realistic methods can be used to combat the problem.

7.4 SUMMARY

The chemical industry is the producer of a multitude of useful products, but there is nearly always a certain amount of the raw material that does not end up in the product. The unused material has to be processed to remove all usable substances and then has to be disposed of so as not to damage the ecological systems described in this chapter. Until now the waste has been disposed of in ways that caused a great deal of damage to air, water, and land. Now it is realized that the environmental damage is real. This chapter discusses this fact and explains just how the damage is caused.

REFERENCES

Braker, W., and A. L. Mossman, *Matheson Gas Data Book,* 5th ed., Matheson Gas Products, E. Rutherford, NJ, 1971.

Crocker, B. B., D. A. Novak, and W. A. Scholle in R. Kirk and O. Othmer, Eds., *Encyclopedia of Chemical Technology*, 3rd Ed., vol. 1, Wiley, New York, 1978, pp. 673–695.

Gertler, A. W, et al., Studies of Sulfur Dioxide and Nitrogen Dioxide Reaction in Haze and Clouds, in J. L. Durham, Ed., *Chemistry of Particles, Fogs, and Rain*, Butterworth, London, 1984.

Foth, H. D., *Fundamentals of Soil Science,* 7th ed., Wiley, New York, 1984, pp. 169–204.

Griffin, H. E., *Ozone and Other Photochemical Oxidants,* Committee on Medical and Biological Effects of Environmental Pollutants, National Research Council, National Academy of Sciences, Washington, DC, 1977.

Haggin, J., *Chem. Eng. News*, **64,** (Mar. 10) 23 (1986).

Innes, W. B., *CHEMTECH*, **14,** 440 (1984).

Lodge, J. R., The Air Environment, in T. E. Larson, Ed., *Cleaning Our Environment—A Chemical Perspective,* American Chemical Society, Washington, DC, 1978, pp. 102–174.

McConnell, J. C., et al., Natural Sources of Atmospheric CO, *Nature,* **233,** 187–188 (1971).

Stern, A. C., R. W. Boubel, D. B. Turner, and D. L. Fox, *Fundamentals of Air Pollution,* 2nd ed., Academic, New York, 1984, pp. 30–31.

Stoker, H. S., and S. L. Seager, *Environmental Chemistry: Air and Water Pollution,* 2nd ed., Scott Foresman, Glenview, IL, 1976.

U.S. Environmental Protection Agency, *Air Quality Criteria for Ozone and Other Photochemical Oxidants*, Environmental Criteria and Assessment Office, Research Triangle Park, North Carolina, 1978.

U.S. Environmental Protection Agency, *1976 National Emissions Report,* Office of Air Quality, Planning and Standards, Research Triangle Park, NC, 1979.

U.S. Environmental Protection Agency, *Air Quality Criteria for Particulate Matter and Sulfur Oxides*, Environmental Criteria and Assessment Office, Office of Research and Development, Research Triangle Park, North Carolina, 1981.

EXERCISE

As part of the trip report in Chapter 1, evaluate the company's operation in the light of the ideas outlined in this chapter.

8

Equipment for Large-Scale Manufacturing

A student is not usually encouraged to think much about the equipment used in his experiments. He is handed a flask, stirring device, and a filter, and is given an experiment to carry out in one or more 2–3 hour periods. We now consider the limitations on his work, particularly as far as apparatus is concerned. Glass is the almost universal flask material.

The first topic to be examined is heat control. In the undergraduate laboratory, the instructions for a typical experiment read to heat the mixture to a stated value and hold the contents at this temperature. The chosen temperature is the one where the reaction proceeds smoothly and where the evolved heat is easily controlled. The experimenter develops skill in judging how fast to heat up and learns not to heat too fast. It has been difficult for some students to develop the skill, as can be seen by looking at most ceilings of chemistry labs. Students have to realize that at a certain temperature, the reaction starts to proceed at a rapid rate, and concurrently there is a brisk heat evolution. Added heat is probably no longer needed. Cooling by water or even ice may be required. Refluxing is a device to hold the mixture at its boiling point. The solvent returning from the condenser has been cooled well below the boiling point. This makes it relatively easy to hold a reaction at this temperature. At other temperatures, the experimenter has pans of hot and cold water to place around the flask as needed for maintenance of a constant temperature.

The next problem the student encounters is to fit the experiment into a three-hour time slot. There is a limited number of experiments that can be completed in this time or that are stable enough to be held at the halfway point for one week.

The university undergraduate is somewhat spoiled as far as effluent is concerned. He does not worry about gas or liquid waste, because it disappears up the hood or down the sink. Some universities are starting to tackle this question in prelab lecture, and it is a key point of emphasis in the author's decision to write this book. A course where the student has to face the difficult problems of properly handling these mixtures is essential for chemistry professionals.

Most of the processes we have been discussing so far have been of the continuous type. These do not resemble lab setups, and were designed for enormous production schedules. They are discussed in Section 8.3. In this chapter we will first discuss batch reactors, which do resemble lab vessels, but are many times larger. They are used when smaller scale, more complicated production is required.

8.1 DESCRIPTION OF A CHEMICAL BATCH REACTOR

Figure 8.1 is a diagram of a typical manufacturing chemical batch reactor. We now discuss it as it relates to the problems that arise in carrying out a variety of processes.

A. Inner Lining of Vessel

The following numbers refer to the diagram in Fig. 8.1. The first factor to be considered is the material of construction. A 2000-L reactor installed in 1986 costs about $120,000. The main concern is that the inside lining of the vessel will not be corroded. The most common linings are glass-lined steel and stainless steel. The glass lining is baked on in the manufacturing process and is the typical choice when an acidic reaction is being carried out. The stainless steel lining (recent vessels have type 316 stainless steel liners) is used when alkaline reactions are carried out. If an acidic reaction is treated to bring the pH to above 9, the contents have to be transferred to a stainless steel vessel. There are several other types of special reactors, such as nickel, brick, and others. The main point is that the person in charge has to be sure that the reaction will not corrode the vessel. He cannot go down to the stockroom as he would do in college and easily obtain a replacement. Think, for example, of the glass chipping off the glass reactor. If this happens there is a special patch, a plug of tantalum, that experience has shown is as unreactive as glass. Putting this patch in place is an expert job. All the equipment that comes in contact with the mixture has to be lined like the vessel walls. This includes the agitator, thermowell, and so on.

B. Heating and Cooling

Heating and cooling are carried out by circulating fluids in the jacket (1A). To heat above room temperature, hot water or steam is used. A temperature of 160°C can be reached with steam under pressure with valves as shown (G, steam in; H trap, J, release). The same jacket is used to introduce cold water for cooling (valves B in, E out) or refrigerated brine (valves A in, F out), which will bring the temperature down to −20°C. (Valves C and D are water and brine overflow.) It is important to note that the jacket extends up almost to the top of the vessel and that the outside of the jacket is insulated over the whole

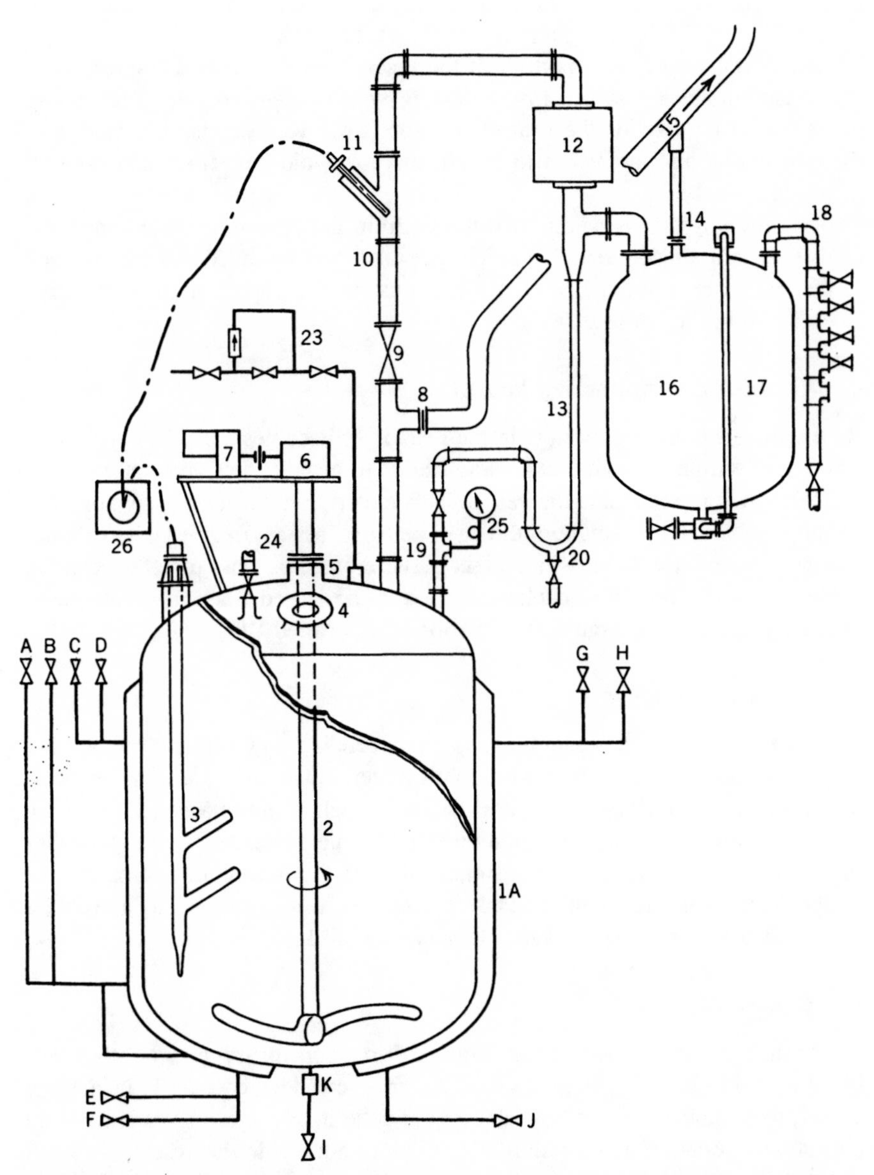

FIGURE 8.1 Diagram of a typical multipurpose chemical reactor. (Source: Ciba-Geigy Corporation, Cranston, R.I.)

exterior. This makes the job of temperature control much more difficult than in the lab. The top of the laboratory reactor is where much heat is lost. The typical lab heater only covers the bottom of the reactor. In the industrial situation, it takes many minutes to switch from a heating cycle to apply cooling. The cooling phase has to remove all the heat of reaction. So a very precise knowledge of the heat evolution picture has to be obtained to avoid dangerous situations in large-scale work.

The reflux system is distinctly different from the laboratory. The vapor line (10) leads to a condenser (12), which takes the thoroughly cooled vapors back to the reactor by another line (13). This prevents the vapor line from flooding as easily as in the lab setup.

C. *Reactor as Separating Vessel*

By inserting a bottom outlet (1) in the reactor, it becomes a separatory funnel. This seems simple enough, but a complication to be taken into account concerns not being able to see inside the vessel. This means that the vessel operator must be informed about the weight of the layer being separated, or at least its volume. The bottom outlet is fitted with a sight glass, a piece of glass pipe between the kettle and the valve. This enables the valve to be closed when the phase separation appears. In the large-scale situation this is called "making a split."

D. *Agitator*

There are many types of agitator available, depending on whether a solid–liquid slurry, two liquid layers, viscous or mobile material, and so on, is being agitated. The agitator in the diagram (2) is called an impeller and is a general-purpose slurry agitator. The setup for agitation in any large vessel requires a stationary baffle (3) to break up the swirling material. The thermocouple to measure the temperature of the reaction is inside the baffle. It is connected to a recorder (26), which also records the vapor temperature at (11).

E. *Cleaning the Vessel*

By the time that the students have completed some chemistry lab work they will be impressed about clean glassware. As they consider chemistry in a large vessel, they should wonder how cleaning is done in a reactor of this sort. In an exceptional situation it is possible for a man to get inside the vessel and wash down the walls. The manhole (4) is the means of entrance. This, however, is not the way that it is normally done. The chemist should have thought about how residues of the reaction can be removed by a series of washes of some sort. Dilute aqueous solutions of acid or base (depending on lining) are usually used. Solvents also would be suitable. The one chosen to start would be the reaction solvent. It could then be added to the solvent to be recovered. Normally, the wash would be raised to the boiling point and some of it distilled into the receiver to wash it out.

All these cleaning hints are obvious. Not so clear is how to make *certain* that the vessel is clean. Realize that there are a lot of crevices (e.g., around gaskets) where crystalline material could have settled. This is solved by refluxing a solvent in the system and evaporating about a liter until only a few milliliters are left. This residue is examined in the infrared for any contaminating substances, and washing with fresh solvent portions is repeated until no residue is found in the IR test.

F. Manhole

The manhole has several other functions, such as adding solid material to the vessel. [Liquid reagents are added by inlet (24).] Also fixed next to the manhole is a point of calibration from which the distance down to the liquid surface in the partly filled vessel can be measured by a calibrated stick. This figure is called the outage. Previous calibration with water and adjustment for specific gravity shows what volume of liquid is present in the vessel.

Another function of the manhole is that it contains a sight glass. This small (3–4-in.), thick glass plate is the only way the operator can see what is going on inside while the reaction is in progress.

The agitator (2) is turned by a motor (7) through a drive shaft (6). A carefully constructed seal (5) permits the agitator to turn while maintaining pressure or vacuum in the reactor.

Figure 8.1 clearly shows that distillation can be carried out to a receiver (16) [(17) is the sight glass to see liquid level]. Reflux is carried out by opening valve (19) to have the condensed liquid return to the reactor. In the latter case, the U-trap (20) becomes an accumulator to remove water when an azeotropic distillation is being performed (see Section 9.1.1).

G. Chemical Manipulations That Are Easy on a Large Scale

There are a number of ways that reaction conditions in large reactors can be readily controlled more carefully than in the lab. Also the kettle is equipped to apply reaction conditions that would need special lab equipment setups. As examples of these, consider that most large manufacturing reactors are fitted with a choice of vacuum systems to be connected to line (18), for maintaining an internal reduced pressure of various values down to 2 mm, depending on the process requirements. The vessel will withstand this and also can be used under pressure up to 90 psi by just closing valves (9) and (19) to the reflux lines, and warming until this pressure is reached. [The pressure gauge is (25).] Also the kettle is capable of cooling the reaction mixture very slowly. This is very important in crystallizing a well-formed product.

H. Safety Features

The nervous laboratory-trained chemist may worry about whether there will be a sudden decomposition inside the huge reactor. Although the previous development work should have reduced this possibility, there is a rupture disk (8) that would break and the pressure of vapors or liquid would be carried through line (15) to a roof container.

Another danger is that of fire, perhaps from a static spark. The danger of this is essentially eliminated by the continual sweeping of nitrogen over the reaction surface through line (23). This elimination of air also reduces the chance of uncontrolled air oxidation.

When the manhole is open, there is a flexible exhaust hose that removes fumes when loading.

8.2 OTHER LARGE-SCALE EQUIPMENT

We now briefly touch on other pieces of equipment. So far, the setups used instead of the flask and separatory funnel have been described.

Trapping Effluent Gases. In place of the laboratory hood, *scrubbers* are used to remove toxic gases. Several such devices are shown in Figure 8.2. The circulating solution traps the effluent gas by neutralizing or some other means. This concentrates in one solution a toxic product that would otherwise be freely disseminated far and wide. Also, the neutralized solution is usually considered suitable to discard. The question of gaseous air pollutants has been discussed in Chapter 7.

Large Filtration Devices. The filter box looks like a Buchner funnel. The solid is retained by porous stones. Two *filtration devices* are shown in Figure 8.3. The centrifuge is a cloth-lined spinning basket (see Figure 8.3*A*). It is used for smaller-volume products such as pharmaceuticals. The filter press (see Figure 8.3*b*,*c*) is for more finely divided material. It consists of a set of large square pieces, either wood or iron. In the center of figure 8.3*b* is shown the frame which collects the solid product. A piece of filter cloth is draped over the top and covers both sides. Each frame is hung on a rack between two plates each filled with a fixed iron grid, (see Figure 8.3*b*), one adapted for slurry transfer and the other for wash liquid transfer. When ready for filtration the plates and frames are pressed together by the machine and a slurry conduit (S) and a wash

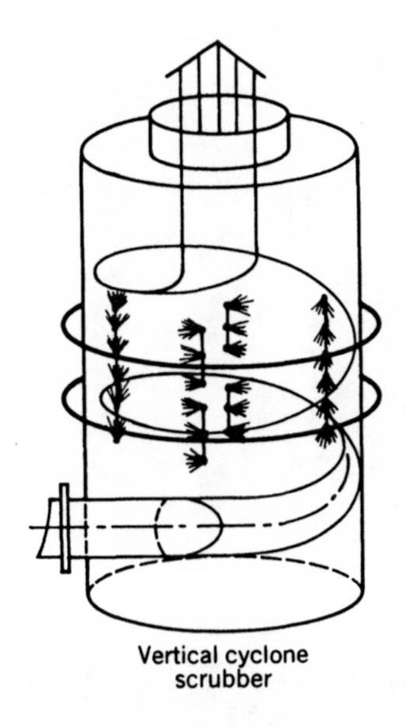

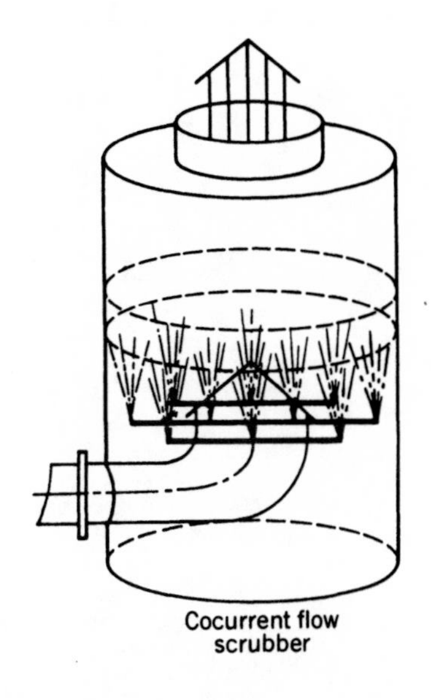

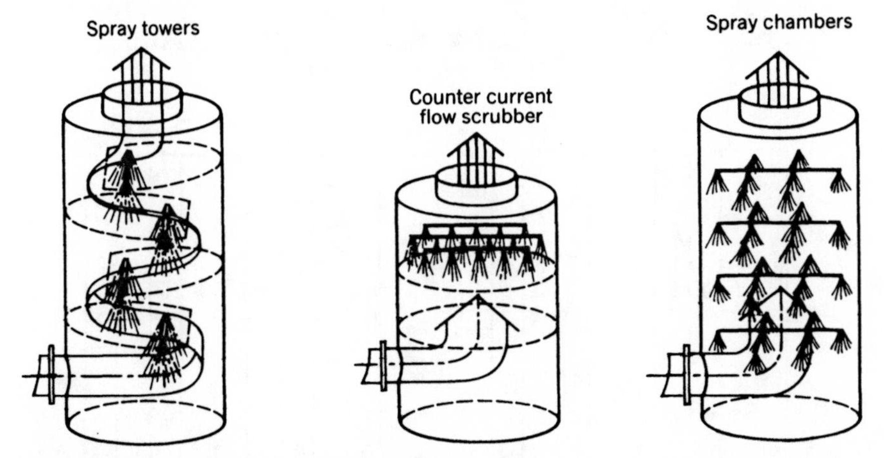

FIGURE 8.2 Scrubbers for trapping toxic effluent gases and particulates. (Source: Ceilcote Corporation.)

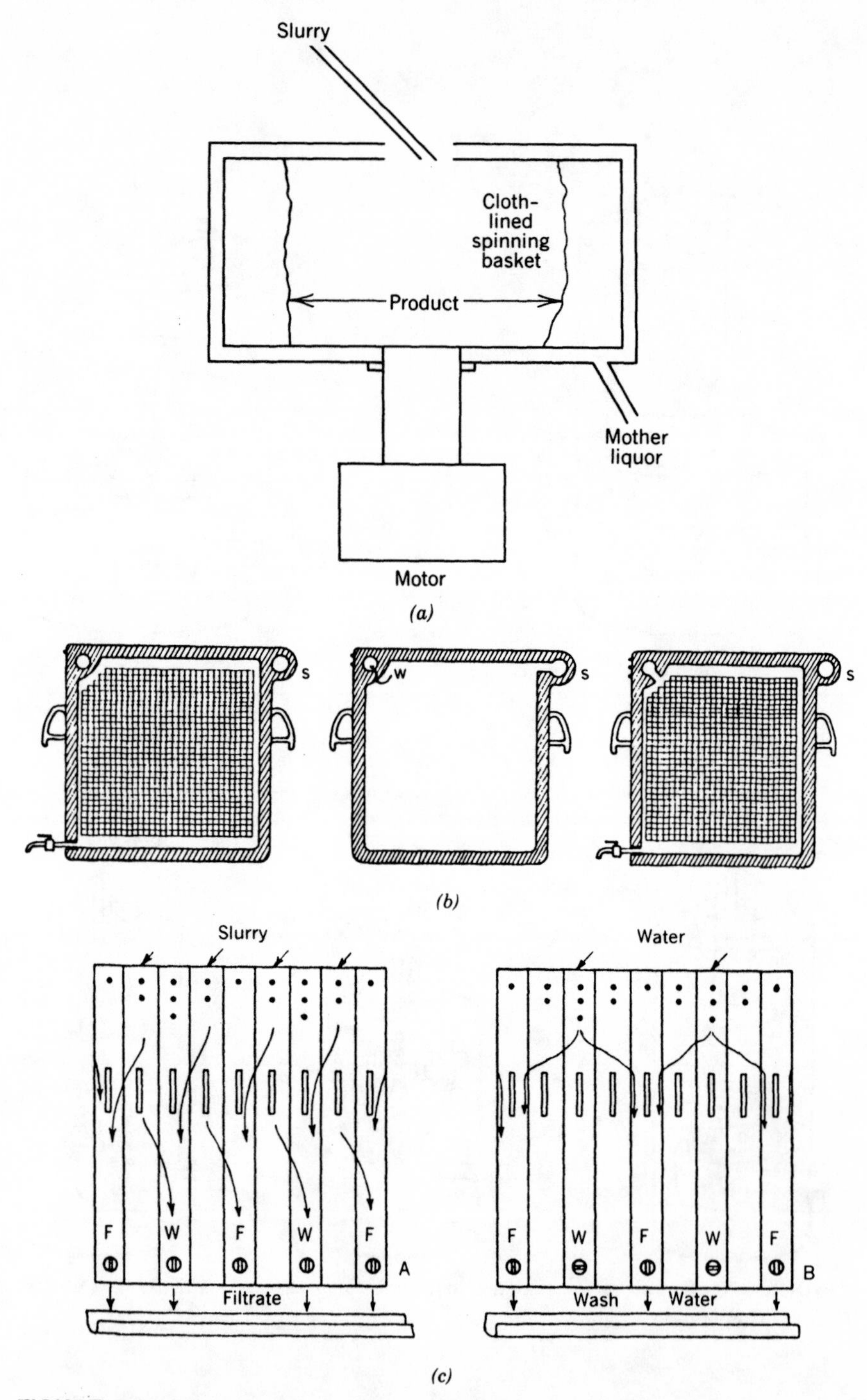

FIGURE 8.3 Filtration devices used in industry. (*a*) Centrifuge, from Ciba Geigy Corp.; (*b*) filter press frames and plates; (*c*) How filtration and washing are carried out in a filter press. (From Riegel, *Chemical Machinery*, Reinhold, New York, 1944, pp. 273–274.)

conduit (W) form (see Figure 8.3*c*). The slurry comes along its conduit and into the frame. The mother liquor runs into the plates and out. The solid collects in the frame. The wash is achieved by leading the solvent through the wash conduit. Its path is through the cake as shown in Figure 8.3*c* and out. Products such as dyestuffs are often isolated in this way.

We have already met with one continuous filter arrangement (Fig. 2.2). This was for the isolation of calcium sulfate from phosphoric acid. Another type of continuous filter has a porous moving belt in a trough. This method is used when filtration takes place at the end of the process; the filtered solid falls continuously into a drum.

Further discussion of large-scale equipment follows.

A. Driers

Two types of driers are shown in Figure 8.4. One is an evacuated box with heated shelves for trays essentially similar to a laboratory drier. The other is a conical drier that rotates as shown and mixes the material during drying. Not shown is an air drier that blows a current of hot air over the solid. This is only useful for water pastes. Each of the vacuum driers has a cooled receiver for the solvent that is removed in the drying process.

8.3 CHOICE BETWEEN BATCH AND CONTINUOUS PROCESSING

There are two ways to make industrial chemicals, by the batch method or by continuous processing. Two articles by Englund (1982) contain excellent introductory reviews of this complex topic.

The batch technique resembles the way the lab experiment is carried out. When one plans a batch, it is necessary to think about the following steps, which take considerable time when large quantities are involved such as when a reactor such as shown in Figure 8.1 is used.

1. Bringing reactants to kettle and loading.
2. Carrying out reaction.
3. Time for analysis and safety checks.
4. Adjusting temperature before transfer.
5. Transferring product from reactor to isolation vessel.
6. Cleaning.

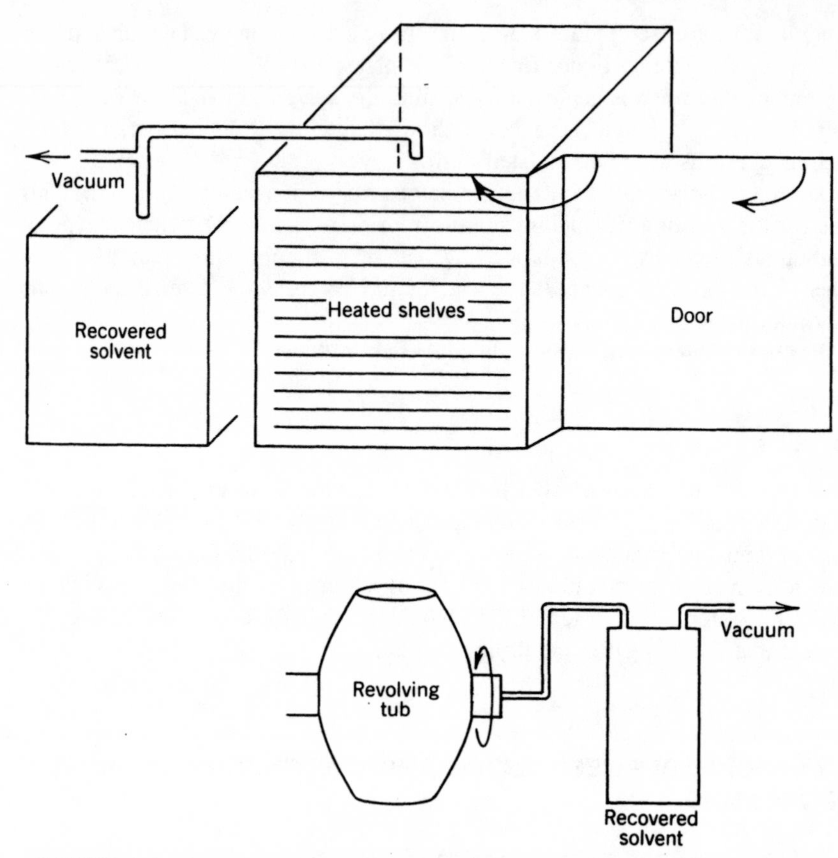

FIGURE 8.4 Chemical driers used in industry. (From Ciba-Geigy Corporation.)

It is clear that in the normal course of events, when a series of preparations are to be made, a considerable amount of time is taken up before and after the reaction.

In continuous processes the feed materials are continuously added to the reactor and the product is continuously withdrawn from the vessel. This eliminates the "dead time" lost when a series of batch runs are performed. Continuous processes clearly require modifications in reactor design.

Figure 8.5 shows a few examples of this type. Vessels *a*, *b*, and *c* are examples where there is "back-mixing" (reactant added to product), and the mixture that results does not resemble a batch mixture. In *b* there would be wall drag giving back-mixing. Reactors *d* and *e* are setups that give a product

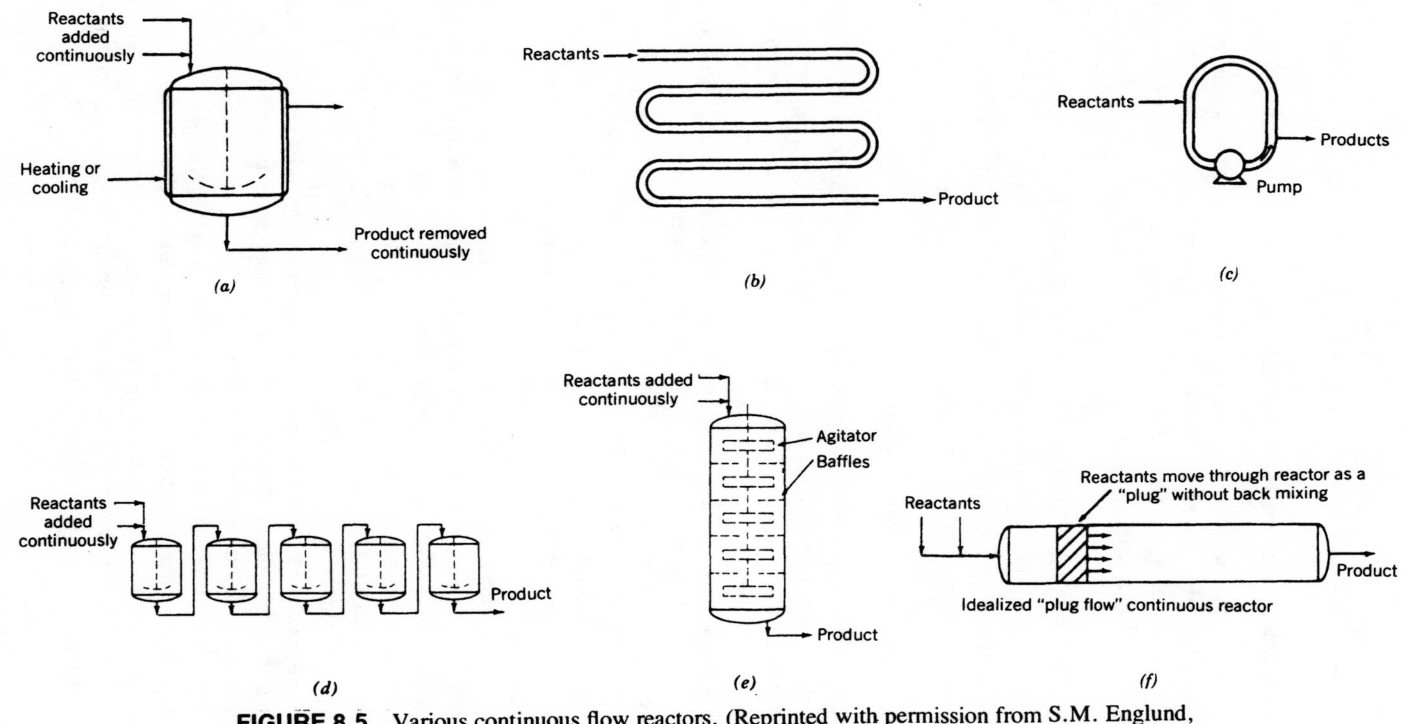

FIGURE 8.5 Various continuous flow reactors. (Reprinted with permission from S.M. Englund, "Chemical Processing—Batch or Continuous, Part I," *J. Chem. Educ.*, **59**, 766–768 (1982).)

mix similar to a batch process even though continuous. Reactor *f* is an idealized plug flow reactor. What emerges would be the same as a finished batch process.

A. Advantages of a Continuous Process

Most large-scale petrochemical processes with just a few steps proceed by continuous procedures. Examples discussed previously are acrylonitrile (Section 4.3.1.B) and vinyl chloride (Section 4.2.2.A). A continuous process once developed is favored because:

1. It is used for making large quantities of one or two products at high rates.
2. It eliminates the time between batches in a series of runs.
3. This process is easier to control and gives a more uniform product.
4. It is easier to design a pressure relief system for a continuous process.

B. Factors That Favor Batch Processes

As pointed out by Englund (1982), batch processes are coming more and more into favor. If a system of kettles can be set up so that a very high occupancy rate is maintained, then this type of process has much in its favor. We now list the advantages of a batch process:

1. Development and scale-up to give a continuous process is more time-consuming than for a batch procedure because of the extra time needed to obtain steady-state conditions. Englund (1982) points out that a 4-hour time cycle batch process gives product in 4 hours. With a continuous process, it might take 10–16 hours before a satisfactory product mixture could be obtained. This is during the process adjustment phase. A great deal of off-standard material will accumulate in this case.

2. Continuous processes are really not suited for mixtures that precipitate on reactor walls. With batch processes, where an automated cleaning can be used after each run, this problem is essentially eliminated.

3. It is possible to increase batch productivity by moving reaction mixtures between vessels and starting other batches or using special kettle arrangements as in the aspirin scheme (Section 8.5).

4. Figure 8.6*a* shows the batch and jacket temperature to be obtained when the maximum reaction rate is obtained with good temperature control for a given cooling fluid temperature in a batch process.

5. To shorten the time between batches, one can add reagents slowly at 25°C and allow the heat of reaction to bring the temperature to the reaction temperature. This saves considerable time between batches, as shown in Figure 8.6*b*–*d*. These diagrams from Englund show the comparison reaction rate graphs for batch, continuous addition batch, and continuous processes.

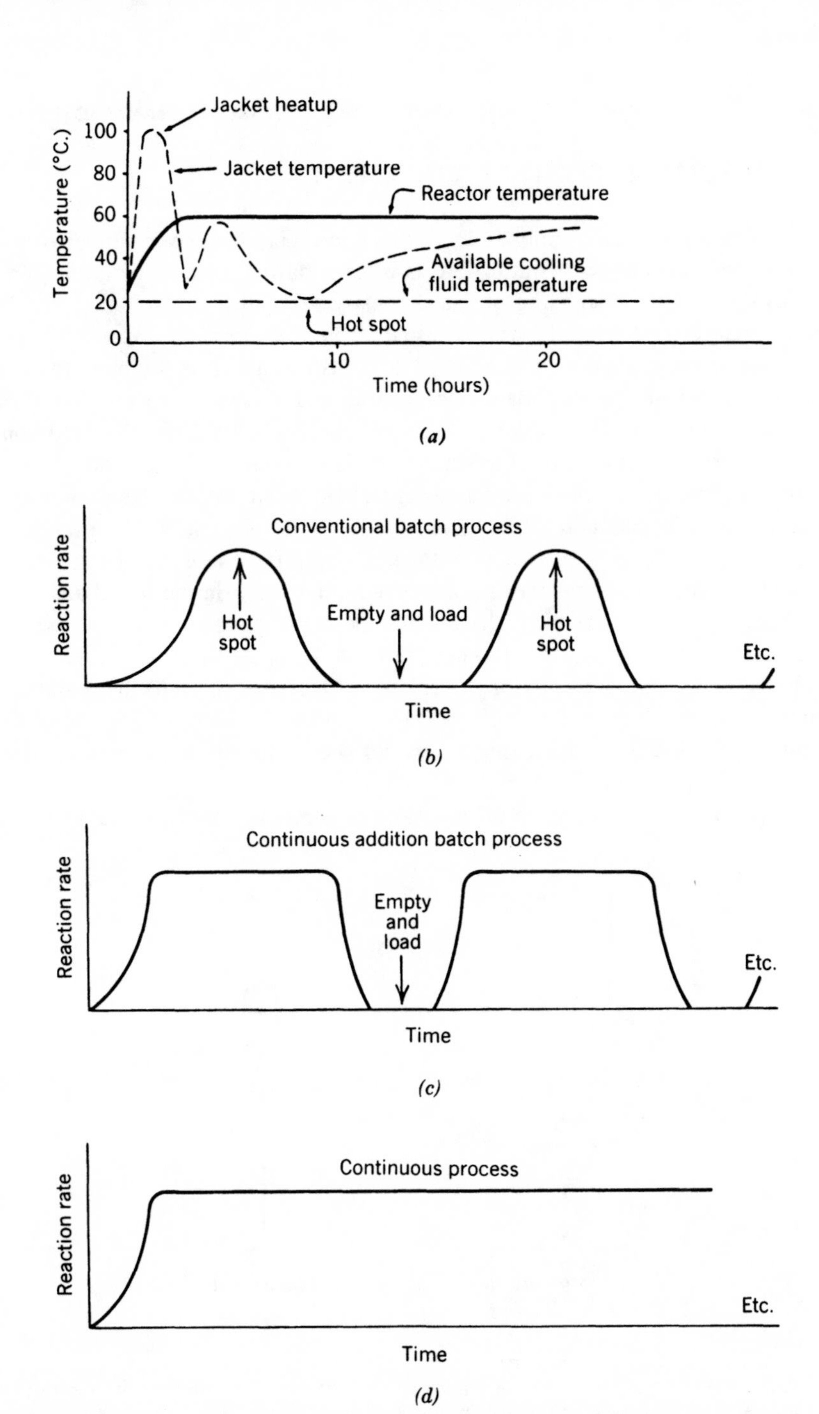

FIGURE 8.6 (*a*) Conventional batch process—maximum reaction rate with good temperature control. (*b*)–(*d*) Productivity graphs for batch and continuous processes. (Reprinted with permission from S.M. Englund, "Chemical Processing—Batch or Continuous, Part I," *J. Chem. Educ.*, **59**, 766–768 (1982); "Part II," **59**, 860–862 (1982).)

8.4 AUTOMATIC CONTROL EQUIPMENT

The laboratory-trained chemist is accustomed to regard his experiment as something under his complete control. He is present during all the operations, and his background understanding of the chemistry makes him aware of any problems. In large-scale work the batches usually run 24 hours a day. New personnel take over in the middle of critical steps as shifts change. It is desirable to have control equipment that automates the opening and closing of valves according to preset instructions. Such devices are called feedback and feedforward controls. An excellent review of automatic controls is given in Clausen and Mattson (1978). Figure 8.7 from this reference gives the basic layout of an automatic control loop. To maintain a beaker of liquid on a hotplate at 50° in a range of 4°, the figure shows a system of automatic controls. These replace the usual control device, an observer with a mercury thermometer in the liquid and with his hand on the heat control. The control loop consists of various devices, in this case a detecting element (taking place of the usual mercury thermometer bulb), a measuring element (for the calibrated mercury column), an automatic controller (replacing the observer), and the final control element (instead of the laboratory rheostat). In this case, one would need a signal-transmitting element (a wire) to replace the observer's arm and hand.

In addition to maintenance of constant temperature, loops can be set up to

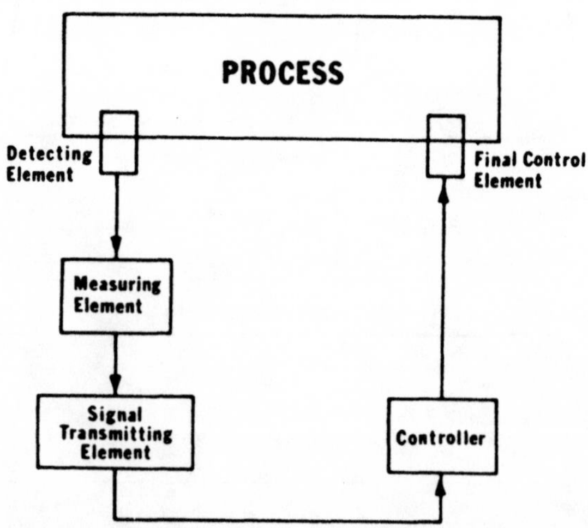

FIGURE 8.7 Basic layout for an automatic control loop. (Reprinted with permission from C.A. Clausen and G. Mattson, *Principles of Industrial Chemistry*, Wiley-Interscience, New York, 1978, p. 232.)

control flow to maintain a given reagent excess, to maintain pressure in a reactor at a stated value, and liquid level in a reactor.

8.4.1 Use of a Set of Kettles to Carry Out a Simple Organic Chemical Preparation

In the description of batch and continuous processes given starting in Section 8.3, it was mentioned that batch processes were not able to produce as much product as continuous ones. Obviously this was because of the time for unloading one reaction and loading the next lot of materials. By careful planning, this "downtime" can be brought to a minimum. The example that follows illustrates how this can be achieved.

We shall use the publication of Norwich Pharmacal (1953) for our example. It briefly covers the preparation of aspirin. It gives few details but does include a quite complete equipment setup. The author has added various details and assumptions to fill in the procedure.

The aim of this exercise is to show how once you have purchased suitably sized vessels, it is possible to keep making batches without stopping. Even though there is only *one* chemical transformation, the process requires 10 vessels of various sizes.

The procedure requires the acetylation of salicylic acid with acetic anhydride using the mother liquor from the previous batch as the medium. The equation is as follows:

$$C_6H_4(OH)(COOH) + (CH_3CO)_2O \rightarrow C_6H_4(OCOCH_3)(COOH) + CH_3COOH$$

This reaction was used as an example for the cost calculation in Chapter 6.

The salicylic acid dissolved in the recovered acetic acid is put in the reactor (see Fig. 8.8), along with the acetic anhydride and some of the mother liquor from the previous batch. We assume that the reaction to make aspirin takes 5 hours from reagent addition to the end of the transfer of the batch to the crystallizer. During this time the batch temperature rises to 90°C. The material is now transferred as a hot solution to the crystallizer.

The crystallization step takes 16 hours, including the time from which the batch starts to go into the vessel, to where the solid and liquid have completely left. During this time the batch temperature drops slowly to 3°C. To avoid a 10-hour wait until the next batch can be started in the synthesis vessel, a set of

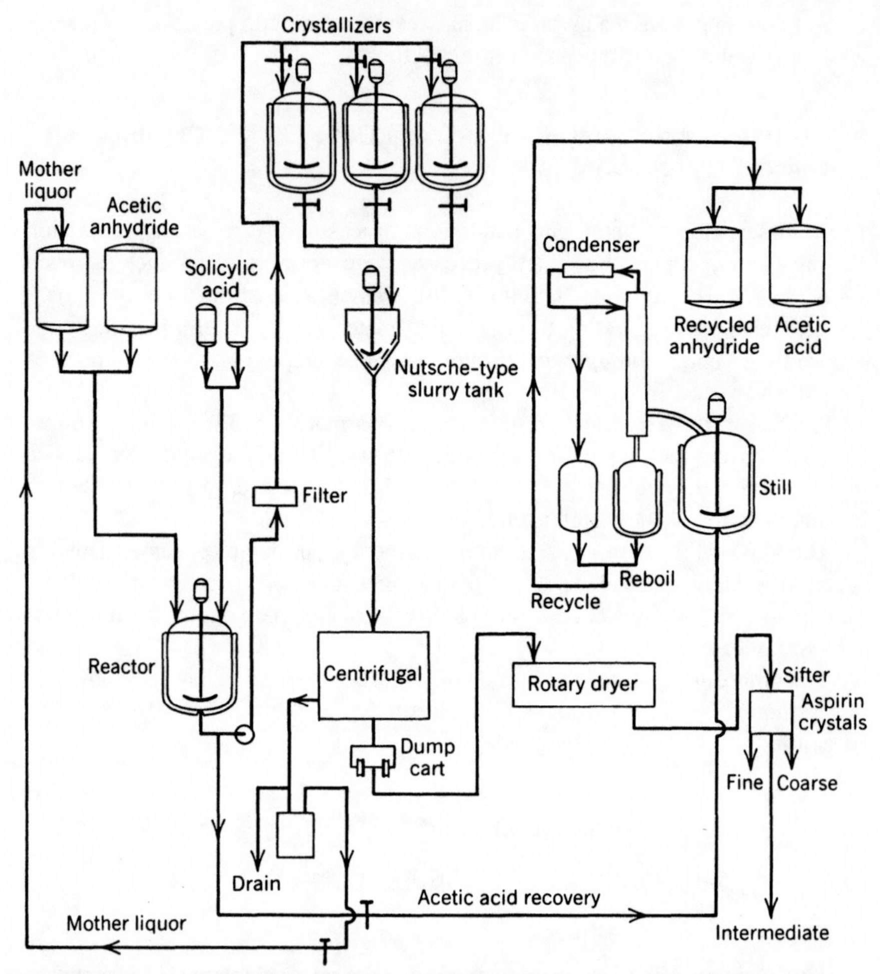

FIGURE 8.8 Kettle reactor layout for aspirin production. (Excerpted from *Chemical Engineering*, June (1953). Copyright 1953 by McGraw-Hill, New York.)

three crystallizers is used. The next synthesis can be started up immediately and placed for crystallization in the second of the crystallizers.

On completing the crystallization step, the batch is transferred to the slurry tank and fed to the centrifuge. We assume that this operation involving the filtration and transfer of the separated crystals to the drier takes 4-5 hours.

The mother liquor not used for the batch is sent to the still for the acetic acid recovery. The acetic acid is separated by distillation.

The crystals from the drier are sifted. Those of intermediate size, about 50% of the batch, are most suitable for tabletting.

There is shown a distillation setup for separation of the acetic acid and the acetic anhydride present in the mother liquor. The acetic anhydride is reused, and the acetic acid can be used in another synthesis after it is used as the reaction solvent.

REFERENCES

Clausen III, C. A., and G. Mattson, *Principles of Industrial Chemistry*, Wiley-Interscience, New York, 1978, pp. 230–259.

Englund, S. *J. Chem. Educ.* **59**, 766, 860 (1982).

Norwich Pharmacal, *Chem. Eng.* 116 (1953).

EXERCISE

In assembling the trip report discussed in Chapter 1, the processes seen should be evaluated according to whether the equipment should be changed in some way. Although you may have little background, this is an opportunity to start thinking about what is involved in assembling the equipment for a process.

9

Basic Background Needed for Large-Scale Operations

A review of distillation, extraction, and crystallization is presented in this chapter. In no sense is this a complete review, but it provides some familiarity with the procedures necessary for the development of a basic chemical manufacturing background.

9.1 DISTILLATION

The distillation process involves boiling a liquid, condensing the vapors, and directing the resulting liquid into another vessel. To carry this out, we need a boiling flask with a vapor column, a thermometer to measure the vapor temperature, a condenser, and a receiver. The liquid is heated to its boiling point, at which point its vapor pressure equals the atmospheric pressure. When the apparatus is at equilibrium, and distillation is occurring, the thermometer indicates the boiling point. As long as the liquid is present and distillation is occurring, the thermometer will indicate a constant value.

The distillation process is used for the separation of two liquids. As an example we consider a mixture of 0.33 mol fraction benzene and 0.67 mol fraction of toluene. By Raoult's law, the vapor pressure of a mixture equals $P_1 + P_2$

Vapor pressure of benzene at 20°C = 75 mm
Vapor pressure of toluene at 20°C = 22 mm

P_1 = vapor pressure due to benzene = $75 \times 0.33 = 25$ mm
P_2 = vapor pressure due to toluene = $22 \times 0.67 = 15$ mm
$P_T = P_1 + P_2 = 25 + 15 = 40$ mm

Dalton's law of partial pressures states that the partial pressure of a gas is proportional to its mole fraction in the mixture.

We have to calculate the composition of the vapor that will give the composition of the condensed liquid from a distillation. Let x_1 and x_2 be the mole fractions of benzene and toluene present in the condensate.

$x_1 = P_1/P_T = 25/40 = 0.63$ mol fraction of benzene
$x_2 = P_2/P_T = 15/40 = 0.38$ mol fraction of toluene

The condensed vapor has almost twice as much benzene in it as does the starting liquid. Now we distill the distillate again. The composition of the condensate is calculated as follows:

Vapor pressure due to benzene = 75 times 0.63 = 47 mm
Vapor pressure due to toluene = 22 times 0.38 = 8.4 mm
P_T = 55.4 mm

Mol fraction benzene = $x_1 = 47/55.4 = 0.85$
Mol fraction toluene = $x_2 = 8.4/55.4 = 0.15$

If we collect the distillate and evaporate again, the vapor will be even richer in benzene. On repeating many times, we obtain essentially pure benzene. This describes what occurs in a fractional distillation. As the material rises through the column, it goes through many evaporation–condensation cycles. The material that reaches the top of the column is richest in the most volatile component. The fractionating column has to be vertical, at equilibrium all the way up, with good mixing of rising vapor and descending liquid at all points.

Figure 9.1 shows this process in graphical form. The boiling points of pure components A and B are T_A and T_B. If an equimolar mixture of these substances X_1 is heated to boiling, the boiling temperature will be T_1 and the composition of the vapor will be X_2. If this is condensed and evaporated again, it will boil at T_2 and the condensed vapor will have the composition X_3. Each equilibration stage is called a theoretical plate.

To evaluate the separation efficiency of a column, one method involves placing a mixture of heptane and methyl cyclohexane in the distillation flask and putting the column on total reflux. Under equilibrium conditions, samples of the distillation flask material and the refluxing material at the top of the column are withdrawn and analyzed. The data previously found for this mixture, plotted as in Figure 9.1, are used to make a new graph of liquid composition vs. the corresponding vapor composition. A connected series of horizontal and vertical lines is drawn between the two composition points indicated by the analysis of the above two samples. The number of steps is called the number of theoretical plates for this distillation column. The larger the number, the better the separation efficiency.

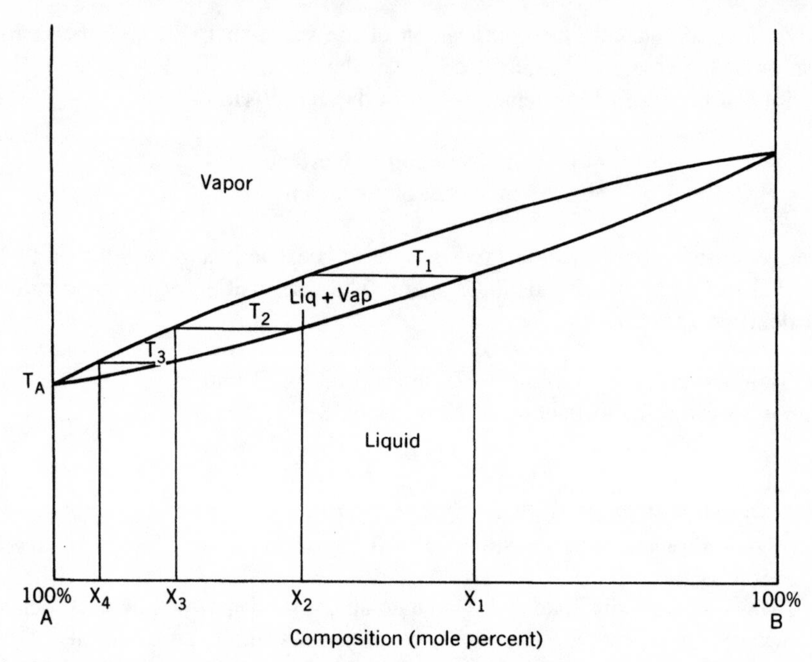

FIGURE 9.1 Distillation of a two-component liquid mixture. Vapor–liquid phase diagram.

To demonstrate the efficiency of a simple and of a fractionating column distillation, Figure 9.2 should be examined. This shows a plot of the distillation temperature against volume of distillate for a simple distillation. The dotted line shows a steadily rising temperature, indicating a continually changing combination of A and B. On the other hand, for a fractional distillation, the solid line shows that the early distillate is essentially pure A and the latter part is made up of mainly B.

The main problem in an industrial fractionation is to pack the column so that equilibration takes place. This is done by a wide variety of devices and materials; however, these materials restrict the distillation rate to various extents. Materials with names such as Rashig rings and Berl saddles are used. In other cases, plates with small holes are fixed at regular intervals in the column.

9.1.1 Azeotropic and Extractive Distillation

In Figure 9.1, a graph is shown for the liquid and vapor compositions for a simple distillation. The use of a fractionating column in such a case will separate the mixture into essentially pure liquids. In Figure 9.3 two vapor–liquid diagrams are shown for mixtures of solvents. These are where the boiling point of

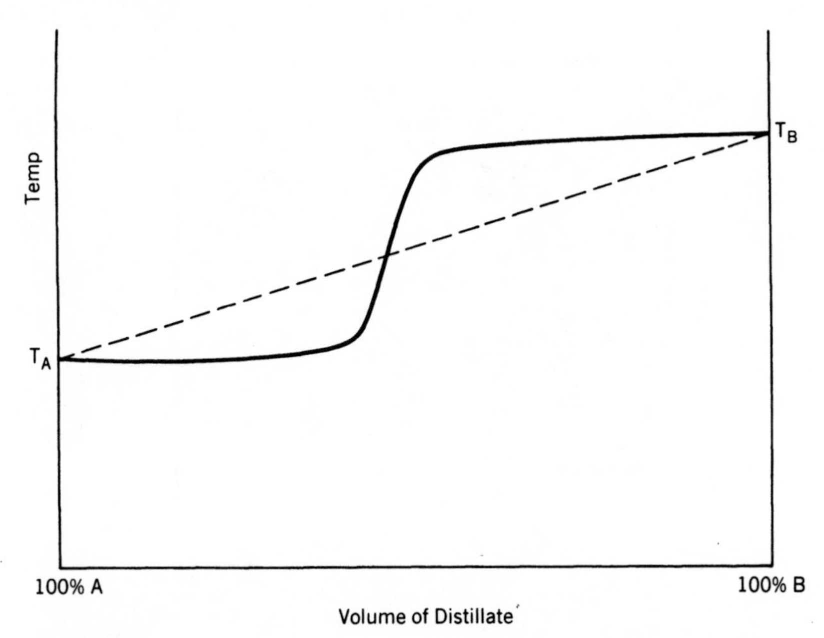

FIGURE 9.2 Distillation curves for a simple and a fractional distillation.

the mixture occurs at lower than or higher than either of the two substances. At the minimum and maximum, the vapor and liquid curves meet. This means that at that composition, that particular mixture will distill unchanged, that is, the composition of the liquid is the same as that of the vapor. For example, with the system with minimum boiling point (Figure 9.3*b*), distillation of a liquid with composition l at temperature t will give a vapor with a composition corresponding to v. Reasoning as in Figure 9.1, using a fractionating column, the lower boiling fraction will have the composition M and the higher boiling fraction will be pure A. Distilling mixture l' in the same way will give the lower boiling fraction with the composition M and the higher boiling will be pure B. When the mixture with the maximum boiling point is studied (Figure 9.3*a*), A or B will distill as lower boiling fractions, and the mixture M will be higher boiling material. The M compositions are called azeotropic mixtures. For the separation of azeotropic mixtures, either of the techniques called extractive distillation or azeotropic distillation is used.

Extractive distillation uses an added solvent of higher boiling point than either of the components. This added solvent should have an attraction for one of the components in the mixture, for example, by hydrogen bonding. As an example, we distill dilute ethanol in a fractionating column, and obtain the azeotropic mixture of 95.6% ethanol and 4.4% water. For the extractive distillation, we transfer this to another column, and use as solvent ethylene glycol. This is added

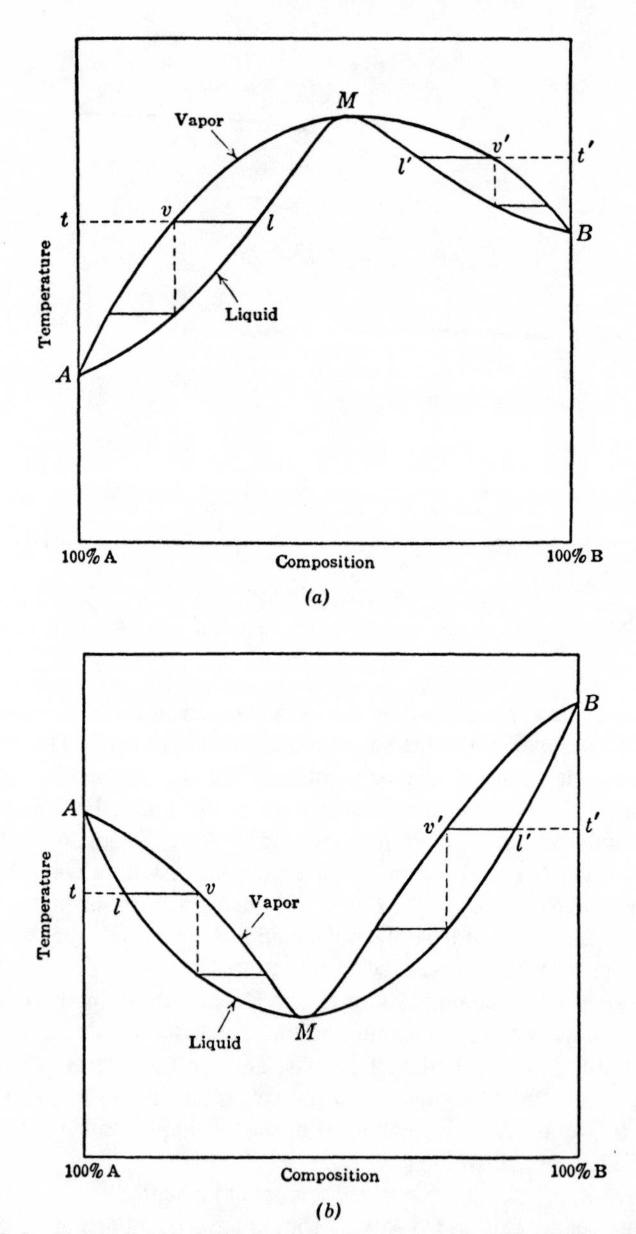

FIGURE 9.3 Vapor–liquid phase diagrams for azeotropic distillations with (*a*) maximum and (*b*) minimum boiling prints.

near the top of the column. When this ternary mixture is distilled, the overhead (lowest boiling component) distilling first is pure ethanol. The less volatile mixture, ethylene glycol and water, is separated by fractionation in another column.

In the *azeotropic distillation* method, a liquid called an entrainer is added to the azeotropic mixture. This material has highly repulsive forces with one of the components (A), that is, there is very low mutual solubility. On distillation, this mixture forms a heterogeneous ternary azeotrope. This is lower boiling than either component and distills over first to be collected in a vessel called an accumulator or decanter where it separates into two phases. The entrainer-rich phase is recycled to the distillation tower. The other phase, mainly the component (A) of the mixture, is processed in another column to recover the entrainer.

With the ethanol–water azeotrope, we can use benzene as an entrainer. We add the benzene at the top of the distillation column in sufficient quantity to remove the water, and distill the ternary azeotrope. The bottom layer of the two phases in the distillate containing most of the water is sent to another column and distilled to the accumulator until the ethanol and benzene is removed as the ternary azeotrope. The remaining liquid is water. When no more heterogeneous distillate is seen, the higher-boiling material remaining in the main distillation vessel is nearly pure ethanol.

An adaption of this method is used in drying water-immiscible solvents which form lower boiling azeotropes with water. The solvent is distilled to an accumulator until no more heterogeneous distillate is seen. The bottom phase (water) is periodically removed and the top returned to the vessel, which now contains dry solvent. See Chapter 8 for the use of this method to prepare dry industrial reactors.

There are two valuable references for this: Molinier (1983) and Prokopakis (1984). This process has been used in Section 4.8.1C.

9.2 EXTRACTION

Distillation is a somewhat wasteful process. There is always material of intermediate boiling point that represents a process loss. Extraction, on the other hand, tends to be more quantitative. When a mixture of substances in solution is brought into contact with an immiscible solvent, one of the materials is usually preferentially extracted. Separation of a mixture into two relatively pure fractions can be accomplished by using two immiscible solvents as follows: The first step is to measure the solubility of each substance S_A and S_B in each solvent. The ratio of the solubilities is called the *distribution coefficient*, and it is an equilibrium constant with a characteristic value for a stated compound in equilibrium with a pair of solvents at a stated temperature.

$$\text{Distribution coefficient} = K_D = S_1/S_2$$

To illustrate the extraction process, consider the following calculation:

A 100-mL aqueous buffer solution contains 1 g of each of compounds A and B. A is an organic acid, and B an organic neutral compound. These materials have the following solubilities.

Compound	Solubility in aqueous buffer (g/100 mL)	Solubility in amyl acetate (g/100 mL)
A	10	2
B	1	10

The distribution coefficients for these materials are (a = amyl acetate; b = aqueous buffer):

$$\text{For A} \quad K_{\text{D}} = \frac{S_{\text{a}}}{S_{\text{b}}} = \frac{2}{10} = 0.2$$

$$\text{For B} \quad K_{\text{D}} = \frac{S_{\text{a}}}{S_{\text{b}}} = \frac{10}{1} = 10$$

Add to the buffer solution 100 ml amyl acetate. Let X be the amount of each material transferred to the amyl acetate phase. This quantity is calculated as follows:

Compound A	Compound B
$K_{\text{D}} = 0.2 = \dfrac{X_{\text{A}}/100}{(1 - X_{\text{A}})/100}$	$K_{\text{D}} = 10 = \dfrac{X_{\text{B}}/100}{(1 - X_{\text{B}})/100}$
$X_{\text{A}} = 0.2(1 - X_{\text{A}})$	$X_{\text{B}} = 10(1 - X_{\text{B}})$
$X_{\text{A}} = 0.167$ g amyl acetate	$X_{\text{B}} = 0.909$ g
$1 - X_{\text{A}} = 0.833$ g buffer	$1 - X_{\text{B}} = 0.091$ g

We are now going to calculate the result where the amyl acetate extraction is carried out on the original buffer solution in four 25-ml portions rather than one 100-ml portion.

First extraction:

$$K_D = 0.2 = \frac{X_A/25}{(1 - X_A)/100} \qquad K_D = 10 = \frac{X_B/25}{(1 - X_B)/100}$$

$$4X_A = 0.2 - 0.2X_A \qquad 4X_B = 10 - 10X_B$$

$X_A = 0.048$ g in amyl acetate $\qquad X_B = 0.714$ g

$1 - X_A = 0.952$ g in buffer $\qquad 1 - X_B = 0.286$ g

Second extraction:

$$K_D = 0.2 = \frac{X_A/25}{(0.952 - X_A)/100} \qquad K_D = 10 = \frac{X_B/25}{(0.286 - X_B)/100}$$

$X_A = 0.045$ g in amyl acetate $\qquad X_B = 0.204$ g

Third extraction:

$X_A = 0.041$ g in amyl acetate $\qquad X_B = 0.059$ g

Fourth extraction:

$X_A = 0.039$ g in amyl acetate $\qquad X_B = 0.016$ g

Adding up the quantities in the four amyl acetate extractions, we obtain 0.173 g for A rather than 0.167 g with the single 100-mL extraction. Also, we find 0.993 g for B, rather than the 0.909 g obtained with one extraction. The following rules of thumb are of use:

With $K \geq 5$, incomplete extraction is obtained; multiple small extractions are more efficient.

With $K < 5$, increase extractant volume.

With $K < 1$, a continuous extraction apparatus is required.

If the amyl acetate were evaporated to dryness, we would have a mixture of 0.993 g, of B with 0.173 g of A. B would be 85.2% pure. We now calculate the increased purity that results if the amyl acetate is backwashed with 50 ml of buffer. Let Y be the amount that goes back to the buffer.

$$K_D = 0.2 = \frac{0.173 - Y_A/100}{Y_A/50} \qquad K_D = 10 = \frac{0.993 - Y_B/100}{Y_B/50}$$

$Y_A = 0.124$ g $\qquad Y_B = 0.047$ g

The amount left in amyl acetate is 0.049 g of A and 0.946 g of B. Again if the amyl acetate were evaporated to dryness, the purity of the compound B would be increased to 95.1%. This exercise shows how the extraction process works.

9.3 CRYSTALLIZATION

It is necessary to understand the basics of crystallization to control crystal size and purity and rate of growth. When a hot, saturated solution is cooled, crystal nuclei have to form in order for the crop to come out of solution.

The nuclei in a carefully filtered hot solution are usually in the form of seeds of a previously purified batch. This is the best way to carry out this process because otherwise the crystals form on the walls of the vessel, especially on any slight imperfections present. It is best to add the seeds just before the point in the cooling cycle at which crystallization can be expected. The seeds should be large enough that they do not have time to dissolve completely.

The rate of cooling has to be controlled to obtain the crystal size desired. Fast cooling will cause the generation of many nuclei, producing a large number of small crystals rather than the growth in size of those already formed. Also, fast cooling will cause inclusion of by-products in the crystals.

The rate of agitation is also a means of controlling crystal size. Fast agitation tends to break up crystals and results in many small crystals. Too slow stirring gives poorly formed crystals, because the neighborhood of the crystal will have become unsaturated in the particular compound present.

It is necessary to find out whether there are other components present that will tend to crystallize with the product.

Whereas the laboratory chemist often tends to use mixtures of solvents to get the necessary selectivity, the industrial chemist cannot often do this. It is quite difficult to recover solvent mixtures economically. The chemist should carry out experiments to obtain a single solvent to achieve the necessary purity of product.

9.4 SUMMARY

This chapter discusses the techniques of distillation, extraction, and crystallization. Each is used very frequently in large-scale work in all kinds of chemistry. In the lab, if adequate purification does not result with one treatment with these techniques, then another extraction, distillation, or crystallization is carried out. In large-scale work, such a situation would not be allowed unless the development laboratory had found it to be absolutely necessary. The correct conditions should be obtained to give satisfaction on the first attempt. It is necessary to become familiar with these all-important procedures.

REFERENCES

Molinier, J., G. Malmary, and J. Constrasti, *J. Chem. Educ.* **60,** 148 (1983).

Prokopakis, G., in R. Kirk and D. Othmer, Eds., *Encyclopedia of Chemical Technology*, 3rd ed., Supplement, Wiley, New York, 1984, p. 145.

EXERCISE

The processes seen on the plant tour discussed in Chapter 1 will doubtless include one or more of the methods reviewed in this chapter. A critical evaluation of the procedures carried out at the plant should be made as part of the trip report.

10

Preparation of Chemical Products from Small Molecules

Up to this point in the book we have concentrated on the basic building blocks of the chemical industry. These small molecules are derived mainly from petroleum, but there are some other sources in nature such as minerals like phosphate rock, oxygen and nitrogen from air, salt from the sea, and organic materials such as carbohydrates and fats.

We have mentioned polymeric products only briefly, as they should be considered in a separate course. More than one-half of all petrochemicals are used to make polymeric materials. The principles developed in this book concerning cost, scale-up, environmental damage and so on, can, of course, be applied to polymers as well as other petrochemicals.

We now give some examples from the pesticide, pharmaceutical and detergent industries to show how these products are made from the small molecule we have been discussing. The small molecule petrochemicals are usually made in continuous processes in large volumes. The products described in this chapter are made in smaller volume and in more complicated processes, and are typically made in batch processes. Further information is given on batch and continuous processes in Section 8.3.

10.1 PESTICIDES

Pesticides are chemical substances that protect crops against a wide variety of insects and other living pests. Chemicals used against weeds are called herbicides. The productivity gains pesticides have produced in agricultural output have been amazing and of fundamental importance in view of the rapid world population increase. We could not go back to a world without these amazing substances. Pesticides have transformed the statistics on human illness and death. For instance, such tragedies as the Irish potato famine of 1840 could never happen again. It was caused by a fungus for which we now have a potent fungicide. It should be noted that monoculture, the practice of growing a single species of plant over an enormous area, has tilted the scale in favor of pests. If a blight of some kind gets established in such a situation, it can spread very quickly because it encounters no barriers. Pesticides and herbicides offer us considerable protection against such disasters.

It has to be stated, however, that environmental problems of a serious nature have been associated with the manufacture and use of some of these substances. Social benefit has to be balanced against overall risk. Agricultural chemicals are a threat to specific parts of the environment in that they are toxic not only to pests but sometimes to some nontarget organisms. This is especially so with the effluent from these processes. Pesticide users and consumers of treated food have to be protected against undesirable side reactions. Contaminated drinking water and air present a distinct hazard to people and wildlife. One purpose of this book is to make the reader aware that there are many ways to control chemical manufacture to make it less hazardous.

10.1.1 Organochlorine Insecticides

We are going to discuss two insecticides—DDT and Lindane. Their use has had a tremendous effect on human health. DDT was used to control the malaria mosquito and was applied to various crops. Lindane has a wide spectrum of activity, including the "soil insect complex" a group of 20 insects. Their synthesis is shown in Figure 10.1.

DDT is synthesized from chlorobenzene (see Table 4.3) and chloral, which is made from the heavy chemicals ethanol (Section 4.2.12) and chlorine (Section 2.6.1.B). Concentrated sulfuric acid (Section 2.4) is also used. Lindane is made by chlorination of benzene (Section 4.1.3.F).

$$C_2H_5OH + 3Cl_2 + 0.5O_2 \rightarrow CCl_3CHO + 3HCl + H_2O$$

$$CCl_3CHO + 2\,C_6H_5Cl \xrightarrow{H_2SO_4} (ClC_6H_4)_2CH{-}CCl_3 \quad \text{(DDT)}$$

$$C_6H_6 + 3Cl_2 \rightarrow C_6H_6Cl_6 \quad \text{(Lindane)}$$

FIGURE 10.1 Equations for the synthesis of DDT and Lindane, chlorinated insecticides.

The residues of DDT are stable in the environment. Indeed, the stability of this material and its apparent nontoxicity to man was considered an advantage. It is now realized that some insects have developed resistance to it, and the populations of successive generations have caused the chlorinated material to get into the food chain. Therefore the use of these materials has been banned in many places. Other products have been developed to take the place of these substances, such as the organophosphorus types.

10.1.2 Organophosphorus Insecticides

Organophosphorus insecticides degrade rapidly in the environment. We discuss two of the many products on the market: methyl parathion and malathion. Their syntheses are shown in Figure 10.2. The ingredients for methyl parathion are phosphorus (see Section 2.2) and the naturally occurring sulfur, methanol (see Section 4.2.5.E), and 4-nitrophenol, made by nitration of phenol (see Section 4.3.4). Malathion uses the same three initial materials plus dimethyl maleate.

The organophosphorus compounds are an outgrowth of World War II nerve gas research. They are more toxic than the organochlorines and require more careful application techniques. They are effective against fewer insects and are more expensive.

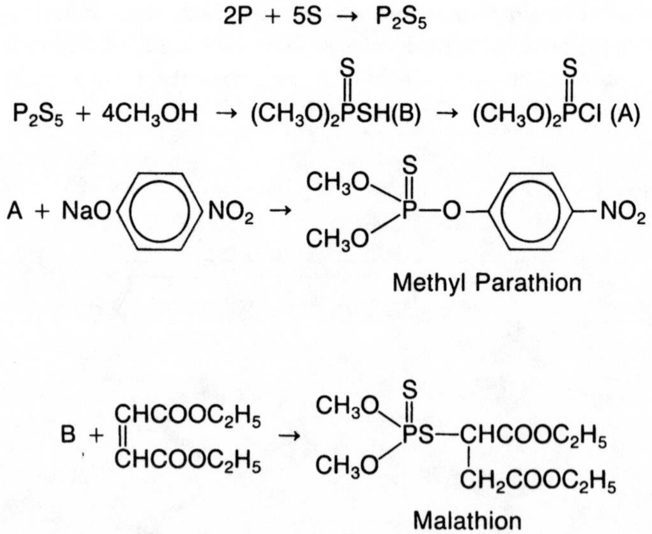

$$2P + 5S \rightarrow P_2S_5$$

$$P_2S_5 + 4CH_3OH \rightarrow (CH_3O)_2P(=S)SH\ (B) \rightarrow (CH_3O)_2P(=S)Cl\ (A)$$

$$A + NaO{-}C_6H_4{-}NO_2 \rightarrow (CH_3O)_2P(=S){-}O{-}C_6H_4{-}NO_2$$

Methyl Parathion

$$B + \begin{matrix} CHCOOC_2H_5 \\ \| \\ CHCOOC_2H_5 \end{matrix} \rightarrow (CH_3O)_2P(=S)S{-}\begin{matrix} CHCOOC_2H_5 \\ | \\ CH_2COOC_2H_5 \end{matrix}$$

Malathion

FIGURE 10.2 Equations for the synthesis of Methyl Parathion and Malathion, organophosphorus insecticides.

We now discuss the toxicity of these materials. They act by phosphorylating the enzyme cholinesterase, which moderates nerve impulse transmission. Depending on the structure of the insecticide, the phosphorylated enzyme has varying stability, resulting in different amounts of persistence in the environment for these materials. Also, their mammalian toxicity is extremely variable. Malathion is very much less toxic to mammals than methyl parathion. Both are toxic to insects. Mammals possess the ability to hydrolyze the succinate ester before it reaches the sites of action. Insects are less able to do this. So we can see that the ability to work out modifications of a basic chemical structure is extremely important in coming up with a satisfactory product. A recent reference with a large number of examples of this sort is by Metcalf (1981).

10.1.3 Carbamate Insecticides

The toxicity of the carbamate compounds is intermediate between that of the organochlorine and the organophosphorus chemicals. They were developed by Geigy of Switzerland. There is a naturally occurring alkaloid physostigmine which has anticholinesterase activity like organophosphates, and a carbamate group. Geigy recognized that carbamates may well have insecticidal activity. A well-known material of this type is the Union Carbide product Sevin, whose structure is given in Figure 10.3. The synthesis of Sevin is also given. The raw

$$CO + Cl_2 \rightarrow COCl_2$$

$$CH_3NH_2 + COCl_2 \rightarrow CH_3N{=}C{=}O$$

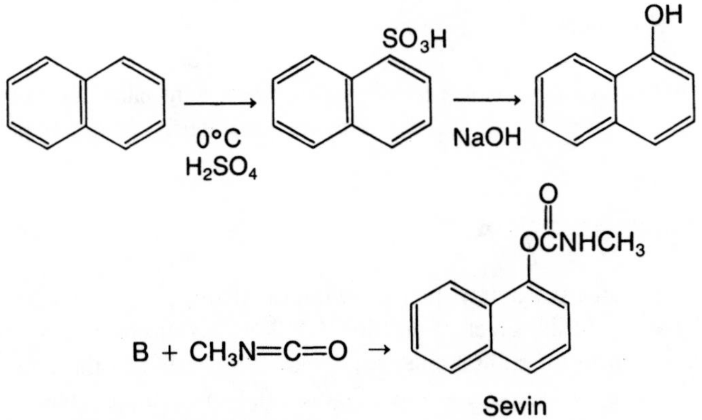

FIGURE 10.3 Equation for the synthesis of Sevin, a carbamate insecticide. (Lambrech, U.S. Patent 2,903,478 (1959, to Union Carbide).)

materials are 1-naphthol and methyl isocyanate. The naphthol is derived from naphthalene, derived from coal tar. The methyl isocyanate is made from methyl amine (see Section 4.8.1.C) and phosgene (made from carbon monoxide (see synthesis gases Section 2.3.2, and chlorine Section 2.6)).

Because these insecticides are rather specific, crops can be protected in an ecologically safe manner. Structural features can be added to these carbamates to protect specific crops. Recent findings, however, show that the carbamate insecticides are rather stable in the soil. The pesticide Temik, a carbamate, has been found to elute into ground water in the southeastern New England area, and this has caused considerable concern about drinking water in wells in the areas sprayed with this material.

10.2 HERBICIDES

For centuries man has kept weeds under control by cultivation, although it has been known that certain chemicals such as sodium arsenite can do this. Such substances often kill all vegetation, however. The new synthetic herbicides are highly specific. The triazine herbicides, for example, tend to kill everything but corn. It has been found that a natural constituent of corn degrades Simazine, a triazine herbicide of Ciba-Geigy, before it can exert its toxic action. There are many kinds of herbicidal action seen, depending on the chemical structure as shown in a book by Buchel (1983):

Contact: kills only that part contacted

Defoliant: causes leaves to drop prematurely

Eradicant: eliminates all vegetation

Systemic herbicides are absorbed by roots or other plant parts and are transported within the plant to tissues that are not necessarily near the point of application.

10.2.1 Triazine Herbicides

The triazine herbicides are photosynthesis inhibitors. They are based on cyanuric chloride, whose synthesis, given in Section 4.8.3.B, is from hydrogen cyanide and chlorine. A family of compounds is available where two of the three chlorines are replaced by an organic amine. Figure 10.4 gives the equation for the most popular of these, Atrazine developed by Ciba-Geigy. The amines used are made as described in Section 4.8.1.C from alcohol and ammonia using a catalyst.

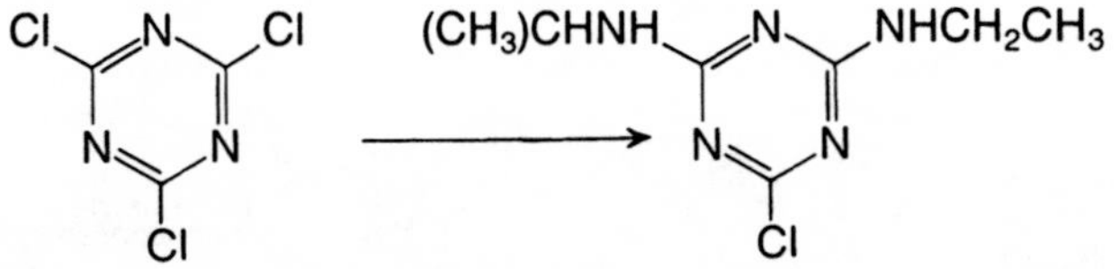

FIGURE 10.4 Equation for the synthesis of Atrazine, a triazine herbicide. (Gysin, Knusli, Swiss patents 342784-5 (1960 to Geigy).)

10.2.2 Phenoxy Herbicides

The phenoxy herbicides were the first synthetic herbicides and were developed by Imperial Chemical Industries of the United Kingdom. They act selectively on broad-leaved weeds and so have been used on many kinds of grain crops such as wheat. They have also been used to get rid of brush along railroad tracks. The raw materials are benzene (Section 4.1.3.F), acetic acid (Section 4.2.4.B), sodium hydroxide, and chlorine (Section 2.6). This product is inexpensive because of the cost of these ingredients. The equation is given in Figure 10.5 for 2,4,5-trichlorophenoxyacetic acid (2,4,5-T).

A development has come with this family of compounds that threatens the further use of these herbicides. As has been mentioned very widely, the analogue 2,4,5-T (see Figure 10.5) was used as a defoliant during the Vietnam War, and there is sufficient evidence to result in the banning of this analogue of the series because of the discovery of the presence of the substance labeled ''Dioxin'' one of a series of related by-products. This is a highly toxic and teratogenic contaminant in both the 2,4,5-T and 2,4,5-trichlorophenol. The structure of Dioxin is given in Figure 10.5, and also the probable mechanism of its formation.

There is another product, pentachlorophenol, whose preparation involves the formation of derivatives of dibenzo-*p*-dioxin. There are two manufacturing processes:

1. Direct liquid phase chlorination of phenol (Boehringer).
2. Hydrolysis of hexachlorocyclohexane at 125–275°C (Dow).

In method 2 *only*, the formation of isomeric dioxins is observed. If the temperature rises above 300°C, a high yield of octachlorodibenzo-*p*-dioxin is seen. Hydrolysis of the chlorine adjacent to the phenolic oxygen is to be expected, leading to dibenzo-*p*-dioxin formation.

This is the most clear-cut example of the need to carry out properly the scale-up of all industrial chemistry processes according to the approach in Chapter 5. There it clearly states that all by-products should be isolated and the structures

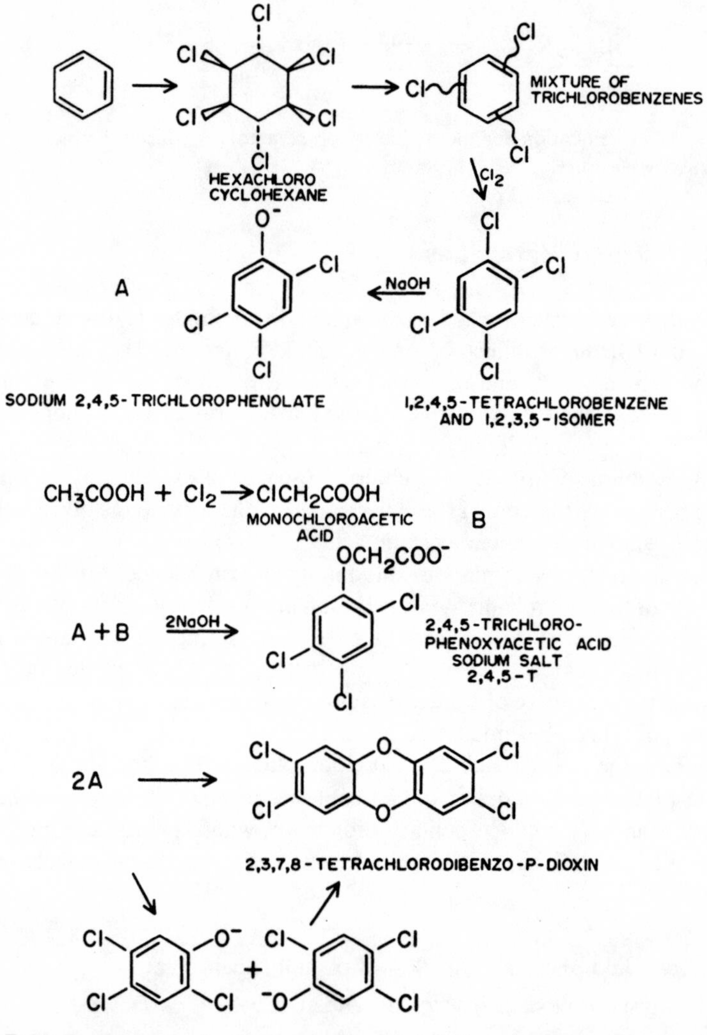

HEXACHLORO CYCLOHEXANE

MIXTURE OF TRICHLOROBENZENES

Cl_2

A

NaOH

SODIUM 2,4,5-TRICHLOROPHENOLATE

1,2,4,5-TETRACHLOROBENZENE AND 1,2,3,5-ISOMER

$CH_3COOH + Cl_2 \rightarrow ClCH_2COOH$

MONOCHLOROACETIC ACID

B

OCH_2COO^-

A + B $\xrightarrow{2NaOH}$

2,4,5-TRICHLORO-PHENOXYACETIC ACID SODIUM SALT 2,4,5-T

2A $\longrightarrow$

2,3,7,8-TETRACHLORODIBENZO-P-DIOXIN

FIGURE 10.5 Equations for the synthesis of 2,4,5-T, a phenoxyacid herbicide. (Pokorny, *J. Am. Chem. Soc.*, **63,** 1768 (1941).)

proved. A decision should be made on what quantity of each of these by-products can be tolerated because of toxicity or other problems.

10.2.3 Amide Herbicides

Propanil is an example of an amide herbicide. These herbicides are extremely selective and have fairly low toxicity to mammals. Like others in the family, Propanil is biodegradable in a single growing season. It is synthesized from 3,4-dichloroaniline, for which the synthesis is shown from chlorobenzene (for preparation , see Table 4.3) and propionic acid (see Section 4.2.4.A). See Figure 10.6. Propionyl chloride, made from the acid and phosphorus trichloride, is the actual reagent used.

10.2.4 Dinitroaniline Herbicides

Trifluralin (see Figure 10.7) is third in volume in herbicides produced after the triazine herbicides and the phenoxyacids. It has a low toxicity. The synthesis has several steps when starting from basic raw materials. 4-Chlorotoluene is prepared from toluene by chlorination with $FeCl_3$ catalyst. This product is converted to the trichloromethyl derivative, which is treated with a derivative of hydrogen fluoride to give the trifluoromethyl derivative. (HF is a by-product of phosphoric acid production, but is evolved as SiF_4; see Section 2.1.1.A) The fluorinated material is converted to the dinitro derivative with nitric acid (see Section 2.5). Reaction with di-*n*-propylamine (see Section 4.8.1.C) as shown completes the trifluralin synthesis.

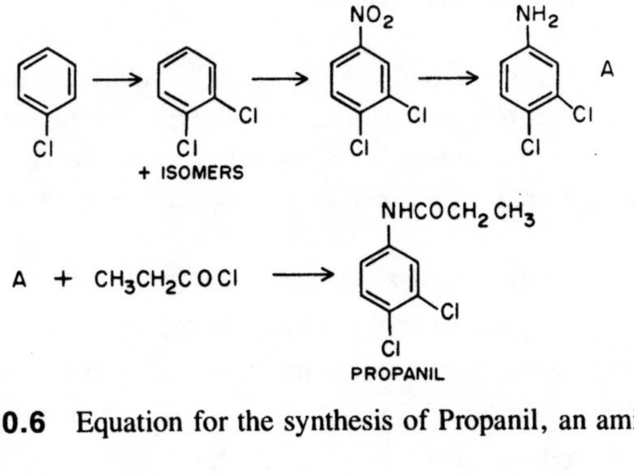

FIGURE 10.6 Equation for the synthesis of Propanil, an amide herbicide.

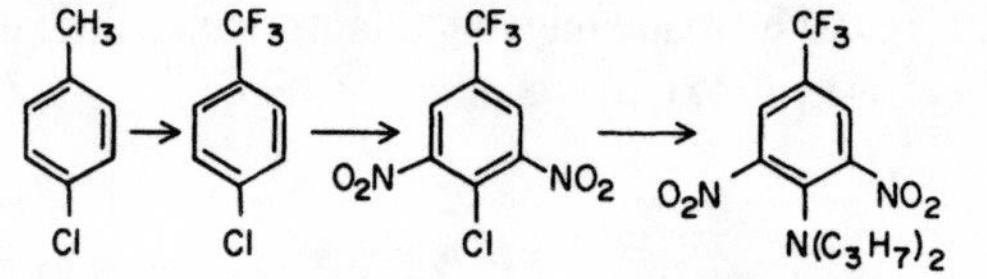

$$\text{CH}_3\text{-C}_6\text{H}_4\text{-Cl} \rightarrow \text{CF}_3\text{-C}_6\text{H}_4\text{-Cl} \longrightarrow \text{CF}_3\text{-C}_6\text{H}_2(\text{NO}_2)_2\text{-Cl} \longrightarrow \text{CF}_3\text{-C}_6\text{H}_2(\text{NO}_2)_2\text{-N(C}_3\text{H}_7)_2$$

TRIFLURALIN

FIGURE 10.7 Equation for the synthesis of Trifluralin, a trifluoromethyl herbicide. (Soper, U.S. Patents 3,111,403 (1960); 3,257,190 (1962) to Lilly.)

10.2.5 Other Agricultural Chemicals

There are a very large number of other herbicides. Some of the chemical families include thiocarbamates and bipyridyl compounds. There are also fungicides such as chlorinated phenols to control plant diseases and rodenticides to get rid of rats, which are estimated to eat 3.5% of the world's grain. Then there are plant growth regulators to control root and blossom development or to control premature dropping of fruit.

10.2.6 Important Factors Raised When Manufacture of a Chemical is Planned

The examples given in Sections 10.2.1–10.2.4 should make the industrial chemist examine the total synthesis of the industrially manufactured materials, and not just the step he has discovered using purchased materials made by another company. It is only then that the impact of the synthesis can be really judged, as far as its use of raw materials and effect on the environment are concerned. It will make the chemist think about whether he or she approaches the synthesis in the really best manner, as far as basic raw materials are concerned.

10.3 SOAPS AND DETERGENTS

Soap, the surfactant known as a cleansing agent since Greek times, is the oldest of many substances that reduce surface tension when dissolved in water. There are four types of surface active agents, or surfactants. These are classified according to the charge held by the surface active molecule in solution:

Anionic: ions in solution bear a negative charge

Cationic: ions in solution bear a positive charge

Ampholytic: ions in solution bear a positive or negative charge, depending on the pH of the solution

Nonionic: dissolved molecule bears no charge

10.3.1 Anionic Detergents

The original soap, from hydrolyzed triglycerides, is an anionic detergent. Since we are focusing on industrial chemistry, the method for carrying out the hydrolysis should be mentioned. This involves autoclave treatment of the fat as a suspension in water with steam under pressure. Neutralization with sodium hydroxide yields soap, the anion molecule.

$$\text{Fat} + H_2O \longrightarrow C_{17}H_{35}COOH \xrightarrow{OH-} C_{17}H_{35}COO^-$$

The next anionic surfactant to be discussed is a sulfonate. This is a more efficient product because it is the salt of a strong acid. An almost neutral water solution will give a product in the anionic form. In the case of the carboxylate salt, a certain amount of hydrolysis of the salt—a salt of a weak acid and a strong base—will occur, which means loss of detergent content.

The most common sulfonate type of detergent is called alkylbenzene sulfonate or ABS. It is prepared by sulfonation of alkylbenzene, whose synthesis is shown in Figure 10.8*a*. The sulfonation step is carried out with sulfuric acid with additional SO_3. This mixture is called oleum (see Section 2.4.2.A).

Figure 10.8*b* shows two procedures for manufacturing the alkylchain. The first gives a branched-chain alkyl product and was the first process used. The oligomerization of isobutylene to a dimer is carried out with an acidic catalyst, yielding a mixture of the double-bond isomers of 2,4,4-trimethyl pentene. The trimer is also prepared in this process. A similar olefin mixture is prepared with propylene as its trimer or tetramer (see Section 4.1.3D). This mixture is used to alkylate benzene (see Figure 10.8*a*). The detergent made from this by sulfonation was an excellent one. The only trouble was that it did not biodegrade in sewage treatment plants or in natural waters, resulting in excessive foaming in both these places. This shows biodegradation was due to the branching of the alkyl groups. A synthesis of a straight-chain olefin of approximately the same chain length was worked out (Fig. 10.8*b*, Part 2). It consisted of a variation of the Ziegler process for polymerization of ethylene, called the Alfen process. The equation is shown in Figure 10.8*b*.

The process takes place in two steps:

1. Ethylene is added to the triethyl aluminum at 100 bar and 90–120°C. The alkyl chain bonded to aluminum grows to the desired length. A mixture of higher trialkyl aluminum derivatives is formed.
2. At 200–300°C and 50 bar, the ethylene oligomers are liberated.

Alkylation of benzene with these straight-chain products followed by sulfonation gives a much more biodegradable product.

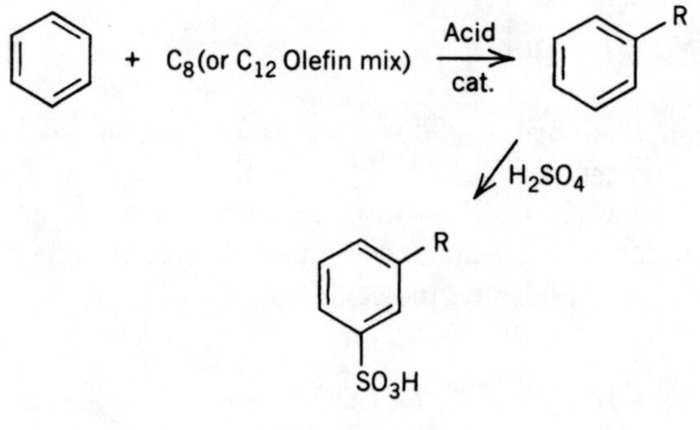

$$C_6H_6 + C_8 \text{(or } C_{12} \text{ Olefin mix)} \xrightarrow[\text{cat.}]{\text{Acid}} C_6H_5R \xrightarrow{H_2SO_4} R\text{-}C_6H_4\text{-}SO_3H$$

(a)

FIGURE 10.8*a* Equation for the synthesis of alkylbenzene sulfonic acid sodium salt, a detergent called ABS. (Huber et al., *J. Am. Oil Chem. Soc.*, **33**, 57 (1956).)

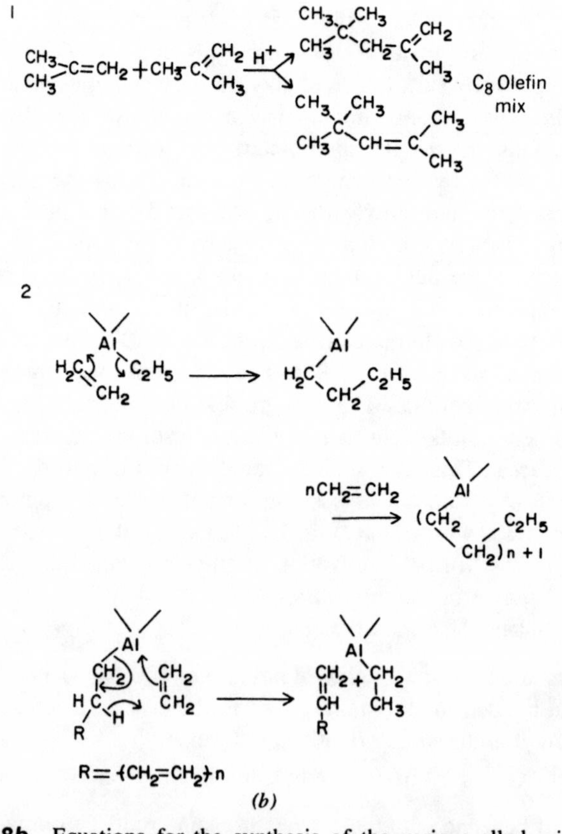

1

$$(CH_3)_2C{=}CH_2 + CH_3\text{-}C({=}CH_2)CH_3 \xrightarrow{H^+} (CH_3)_3C\text{-}CH_2\text{-}C({=}CH_2)CH_3 \;;\; (CH_3)_3C\text{-}CH{=}C(CH_3)_2$$

C_8 Olefin mix

2

$$Al\text{-}C_2H_5 + H_2C{=}CH_2 \longrightarrow Al\text{-}CH_2\text{-}CH_2\text{-}C_2H_5$$

$$\xrightarrow{nCH_2{=}CH_2} Al\text{-}(CH_2\text{-}CH_2)_{n+1}\text{-}C_2H_5$$

$$Al\text{-}CH_2\text{-}CH(H)R + CH_2{=}CH_2 \longrightarrow CH_2{=}CH\text{-}R + Al\text{-}CH_2\text{-}CH_3$$

$R{=}\text{-}(CH_2{=}CH_2)\text{-}_n$

(b)

FIGURE 10.8*b* Equations for the synthesis of the various alkyl mixtures used for making ABS detergent. (Weissermel and Arpe, *Industrial Organic Chemistry*, Verlag Chemie, Weinheim, 1978, pp. 65, 68.)

Another problem with the new detergents was that they did not clean the new synthetic fabrics well. The first approach was to use sodium carbonate. The best solution was to use sodium tripolyphosphate, which promoted the cleaning action and softened hard water by chelation of heavy metal ions. However, another problem surfaced: The effluent waters now tended to support heavy growths of algae. It was considered that this was due to the phosphate content from the detergent as well as the other nutrients. It is now mandatory that waste treatment plants consider treatment to reduce phosphate, especially where discharge of treated effluent is into inland waters (see Section 3.2.4.C).

10.4 PHARMACEUTICALS

The pharmaceutical industry produces curative agents in about 25 branches of medicine. As everyone knows, truly spectacular products have often been created. We have already referred to one, penicillin, a fermentation product. There are a great many other antibiotics, many of which are fermentation products.

As an example of a synthetic product that represents a remarkable synthesis achievement and also a successfully scaled-up commercial preparation, we now look at the preparation of vitamin A acetate, shown in Figure 10.9.

The synthesis is from the work of the group with the BASF corporation. The work was initially described by Pommer (1960) and Rief and Grassner (1973). Later work was carried out to adapt the synthesis so that it would use materials already on hand at BASF having a petrochemical basis. The synthesis is outlined by BASF (1985). The synthesis uses the following chemical substances, that have already been discussed in this book, in the preparation of the huge vitamin A molecule:

isobutene
formaldehyde
acetone
vinyl chloride
acetylene
acetic acid
carbon monoxide/hydrogen mixture (syngas)

Triphenyl phosphine is also used. This is prepared from phenyl magnesium chloride (made from chlorobenzene and magnesium) and phosphorus trichloride (made from red phosphorus and dry chlorine).

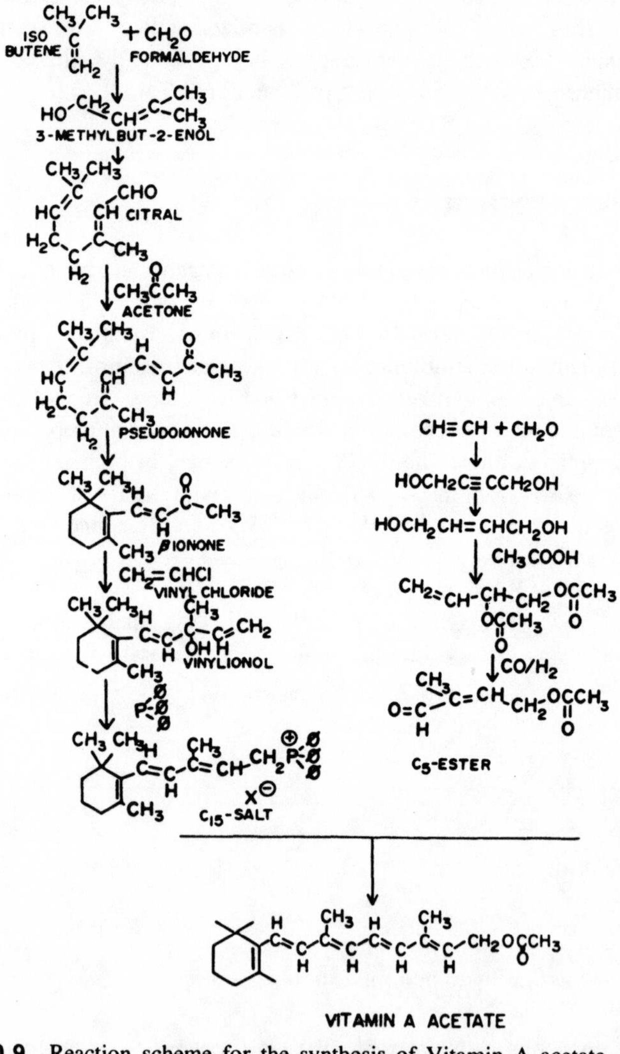

ISO BUTENE + CH_2O FORMALDEHYDE

3-METHYLBUT-2-ENOL

CITRAL

CH_3COCH_3 ACETONE

PSEUDOIONONE

β IONONE

$CH_2=CHCl$ VINYL CHLORIDE

VINYLIONOL

C_{15}-SALT

$CH \equiv CH + CH_2O$

$HOCH_2C \equiv CCH_2OH$

$HOCH_2CH=CHCH_2OH$

CH_3COOH

CO/H_2

C_5-ESTER

VITAMIN A ACETATE

FIGURE 10.9 Reaction scheme for the synthesis of Vitamin A acetate. (Reprinted with permission from *From Research to Practical Experience*, BASF Aktiengesellschaft, Ludwigshafen, 1985.)

This synthesis illustrates a basic theme of this book:

Use raw materials made in large quantities, which are usually made in high yield and consequently at lowest cost.

10.5 SUMMARY

In this chapter petrochemicals and other raw materials are used to prepare useful products from several parts of the chemical industry listed in Chapter 1. The examples use batch or continuous processes. The discussion emphasized the importance of considering the basic raw materials used for the synthesis.

REFERENCES

BASF, *From Research and Practical Experience*, BASF, Ludwigshafen, 1985, p. 3.

Buchel, K. H., *Chemistry of the Pesticides*, Wiley, New York, 1983.

Metcalf, R. L., in R. Kirk and D. Othmer, Eds., *Encyclopedia of Chemical Technology*, 3rd Ed., Vol. 13, 1981, pp. 413–485.

Pommer, H., *Angew. Chemie*, 72–811 (1960).

Reif, W., and H. Grassner, *Chemie. Ing. Techn.* **45,** 10 (1973).

EXERCISE

Take any three important materials, for example, a pharmaceutical or an agricultural chemical, not covered in this chapter and trace the synthesis back to the basic raw materials. This discussion should use the most up-to-date literature. A comparison with an earlier synthesis, again going back to basic raw materials, should be made.

11

Industrial Catalysis

It has been stated that about 70% of all industrial processes involve catalyst use. A review of how industrial chemists look at catalysts is presented in this chapter. The field of catalysis has gone far beyond the strictly empirical approach of a few years ago. We cover briefly the areas of homogeneous and heterogeneous catalysis, emphasizing industrial aspects.

First we must define two catalytic process types, fixed bed and fluidized bed. In fixed beds, the catalyst is held in place by grids or nets on shelves or in tubes. The vapors flow by. In fluidized beds, the catalyst has a specific physical form through which vapors move. This causes the solid catalyst to move about like a liquid.

11.1 HETEROGENEOUS CATALYSIS

Satterfield (1981) gives an excellent overview of industrial heterogeneous catalysis. He discusses the basic idea of adsorption of reactants on the catalyst surface, reaction and product desorption, and the physical aspects of the process (geometry of the active sites at specific points on the catalyst surfaces), how these are affected by particle size, and other things. The active sites are charge carriers, giving or taking electrons. He also discusses this process from the chemical aspect, mentioning that the bond strength between the reactants and catalyst should not be too weak or too small for the maximum rate.

We now discuss how an industrial chemist thinks about this type of catalysis. The usual temperature range is 20–500°C. Below this range, reactions are too slow, and when the temperature is greater than 500° selectivity is not adequate. A few reactions that we discussed earlier are run above 500°. These are considered to have unusually stable products. These include:

1. $NH_3 + O_2 \xrightarrow[\text{Pt-10\% Rh} \quad 920°C]{} NO_x$ (see Section 2.5).
2. $CH_4 + O_2 + NH_3 \xrightarrow[\text{Pt-10\% Rh gauze} \quad 1000\text{–}1200°C]{} HCN$ (see Section 4.8.3.B).
3. $CH_3OH + O_2 \xrightarrow[\text{Ag} \quad 600°C]{} HCHO$ (see Section 4.8.1.A).

4. $CH_4 + H_2O \xrightarrow[\text{Ni } 700\text{–}800°C]{} CO + 3H_2$ (see Section 2.3.1.A).

5. Automobile catalytic converters which oxidize waste hydrocarbons and reduce NO_x operate at high temperatures also (see Section 11.1.2.A).

The industrial chemist looks at a catalytic process, examining how the rate and direction of the process under study is affected by

Changes in catalyst composition
Changes in feed composition
Changes in pressure and temperature
Changes in degree of recycle
Changes in contact time and aging

In these studies the chemist works with a mechanically stable, selective, and highly active catalyst. These studies take place during the process development stage where the best process is being developed.

On the other hand, to decide on the mechanism of reaction, the study of material absorbed on single-catalyst crystals is carried out.

11.1.1 Selecting Catalysts

We are getting to the stage where we are no longer pulling a succession of compounds off the shelf to try in the process. We look for substances having strong ionic or metallic bonding (strong interatomic fields) to have catalytic activity. (See Table 11.1.)

Satterfield (1981) has grouped the catalyst types, relating the chemical description of the material to the general sort of chemical reaction it catalyzes. In all cases the substances that act as catalysts have metallic or ionic bonds, and these hold on to their structures under reaction conditions.

Naming Catalysts. Satterfield points out that we now realize that catalysts do react according to the principles of chemistry. In this book the author has given examples of mechanisms illustrating this point. In spite of this, however, the reader should realize that the problem is to describe the active elements and the support to specify the form in which the element may exist in the catalyst as manufactured or under reaction conditions. An example of this is in Section 2.4.2.B, dealing with the oxidation of SO_2 to SO_3 using vanadium pentoxide promoted with an alkali metal sulfate. Satterfield gives several other examples of this problem in Table 11.1.

TABLE 11.1 Heterogeneous Catalysts of Industrial Importance

Catalyst Type	Catalyst	Section
Hydrogenation	Fe promoted with KOH,	
ammonia synthesis	and with Al_2O_3,SiO_2 and	
$N_2 + 3H_2 \rightarrow 2NH_3$	MgO	2.3.1.C
Dehydrogenation	Fe_2O_3 promoted with	
Ethyl benzene to styrene	$(Cr_2O_3 + K_2CO_3)$	4.2.10
Butenes to butadiene	Same as above	4.4
$CH_4 + H_2O$ to $CO + 3H_2$		
Partial oxidation	Supported Ni	2.3.2
$CH_3CH{=}CH_2 + O_2 + NH_3$ to acrylonitrile	$Bi_2O_3 \cdot nMoO_3$	4.3.1
Chemisorbed oxygen	+ other components	
Ethylene to ethylene oxide	Supported silver	4.2.8
CH_3OH to HCHO,	Silver	4.2.9.A
Strong oxidizing catalysts	or Fe_2O_3–MoO_3	
Complete oxidation of hydrocarbons—auto exhaust	Pt–Pd pellet on monolith support	11.1.2.A
Butene + O_2 to maleic anhydride	Supported V_2O_5	4.7.2.A
SO_2 to SO_3	$V_2O_5 + K_2SO_4$	2.4.2.A
Acidic catalysts		
Catalytic cracking	Zeolite (molecular sieve) in silica–Al matrix	4.1.3B
Catalytic reforming	Pt–Re on acidified alumina	4.1.3.F
CH_3OH to gasoline	Mobil ZSM-5 zeolite	4.8.2.A
Reactions of syngas		
$CO + 2H_2$ to CH_3OH	Highly active Cu	4.2.5.A
$CO + 3H_2$ to $CH_4 + H_2$ (methanation)	Supported Ni	2.3.2

A. *Selecting Catalysts for Various Processes*

The metallic catalysts for hydrogenation and dehydrogenation absorb hydrogen readily. The bond is rather weak. The group 8 (Fe, Co, and Ni) and the platinum metals are used here. (See the ammonia preparation, Table 11.1).

The second group of metals are used in oxidations. These are the platinum group (Ru, Rh, Pd, Os, Ir, and Pt). These metals chemisorb oxygen on their surfaces. Most other metals are converted to their oxides throughout the sample. The chemisorption bond with oxygen is stronger than with hydrogen, which results in the catalytic oxidations requiring a higher temperature than the hydrogenations.

There are two groups of oxide catalysts. One is for partial oxidation reactions and the other for dehydrogenations. In the first case we have an ionic structure of a sort where the oxygen atoms can be transferred readily into and away from the catalyst. Examples of these are complexes containing molybdates of several metals used for ammoxidation of propylene to acrylonitrile (see Section 4.3.1.B), iron molybdate for conversion of methanol to formaldehyde (see Section 4.8.1.A), and vanadium pentoxide for oxidation of naphthalene or *o*-xylene to phthalic anhydride (see Section 4.7.2).

In the second type the oxygen is more tightly bound to the catalyst. The oxide must not be reduced to the metal by the eliminated hydrogen at the reaction temperature. See ethyl benzene to styrene using Fe_2O_3 with Cr_2O_3.

Acid-catalyzed reactions are brought about by various acidic solids. In these cases two or more atoms of the element are linked by oxygen atoms, as in the zeolites on an aluminosilicate matrix used in catalytic cracking (see Section 4.1.3.A), the zeolites used in the methanol to gasoline process (see Section 4.8.2.A), and the acid-type ion-exchange resin used in hydration of propylene to isopropyl alcohol (see Section 4.3.5).

11.1.2 Catalyst Selectivity

A review by Sachtler (1983) on this topic lists four headings under which to consider the subject of selectivity. This topic is important when there are several thermodynamically possible products from some reacting materials. Sachtler gives the example of the oxidation of ethylene to yield:

1. Ethylene oxide using a silver-based catalyst (see Section 4.2.8).
2. Acetaldehyde using the Wacker catalysis by palladium and copper(II) ions (see Section 4.2.3.B).
3. Carbon dioxide and water by combustion. This process is catalyzed by platinum as in the auto afterburner.

To have just selected processes proceed selectively to the exclusion of others, we must understand the chemical reactions to be catalytically favored, and also those which we do not want to occur. A recently reviewed relevant case involves the auto catalytic converter discussed below.

A. *Auto Catalytic Converters*

Reviews by Taylor (1984) and Hegedus and Gumbleton (1980) cover this topic. The aim with these devices is to treat the auto exhaust fumes to convert carbon monoxide and hydrocarbons (HC) to carbon dioxide and water, and nitric oxide to nitrogen. Exhaust emission characteristics vary according to air/fuel ratio

(*A*/*F*). The untreated emissions are made up of paraffins, olefins, aromatics, water, carbon monoxide, and 20 ppm sulfur dioxide. The stoichiometric *A*/*F* ratio is 14.6, and this is the optimum for conversion of CO and NO. A catalyst mixture has been found to convert CO to CO_2 and also NO to N_2. This is a mixture of platinum, palladium, and rhenium, and is called a three-way catalyst system. When the system is at 500°C, the typical auto exhaust temperature, the desired products are thermodynamically favored, and reactions (1)–(5) below proceed, as well as some (6).

The chemical reactions important in this work are:

$$CO + 0.5O_2 \rightarrow CO_2 \tag{1}$$

$$HC + O_2 \rightarrow H_2O + CO_2 \tag{2}$$

$$H_2 + 0.5O_2 \rightarrow H_2O \tag{3}$$

$$NO + CO \rightarrow 0.5N_2 + CO_2 \tag{4}$$

$$NO + H_2 \rightarrow 0.5N_2 + H_2O \tag{5}$$

$$HC + NO \rightarrow N_2 + H_2O + CO_2 \tag{6}$$

$$NO + 2.5H_2 \rightarrow NH_3 + H_2O \tag{7}$$

$$CO + H_2O \rightarrow CO_2 + H_2 \tag{8}$$

$$HC + H_2O \rightarrow CO + CO_2 + H_2 \tag{9}$$

With a rich *A*/*F* ratio, less air than the stoichiometric values ($A/F = 14/6$) the oxidizing agents are in short supply. HC, CO, and H_2 compete for NO and O_2. With a lean *A*/*F* ratio (reducing agents in short supply), NO and O_2 compete for HC, CO, and H_2. In other words, the three-way catalyst promotes reactions (1) and (2) rather than (3), and (4), (5), and (6) rather than (3). The review by Taylor (1984) points out that close to the stochiometric values of *A*/*F*, the desired reactions all proceed satisfactorily. The catalyst favors the conversion of NO to N_2 rather than (7), the formation of ammonia. Rh is relatively insensitive to SO_2 poisoning. This is the main reason for its use. There is the likelihood of forming NH_3 when the emission is completely free of SO_2. With the typical exhaust there is essentially no NH_3.

The *A*/*F* ratio has to be continually adjusted as the content of the emissions changes.

B. *Chemical Requirements Involved in Choosing a Catalyst*

Having shown by examples how precisely catalysts can be applied in a complex system, we now examine Sachtler's approach to describing a selective catalytic mixture for any particular case. Four chemically defined requirements are described which would be the basis for experiments to come up with the best catalyst for any reacting system.

Bond Strength Requirement. A heterogeneously catalyzed reaction is carried out with a series of catalysts. The energy of activation is plotted against the strength of the chemisorption bond. This plot passes through a minimum. This is the point that chemically describes the most suitable catalyst for the reaction. No chemisorption will take place if the bond is too weak. The desorption step is too slow if the bond is too strong.

Coordination Requirement. For some catalyzed reactions to take place, more than one coordination site is needed per atom. Sachtler (1983) uses as an example of this the equilibration of H_2 with D_2. The following mechanism requires three coordination positions on the catalyst that can be occupied by H.

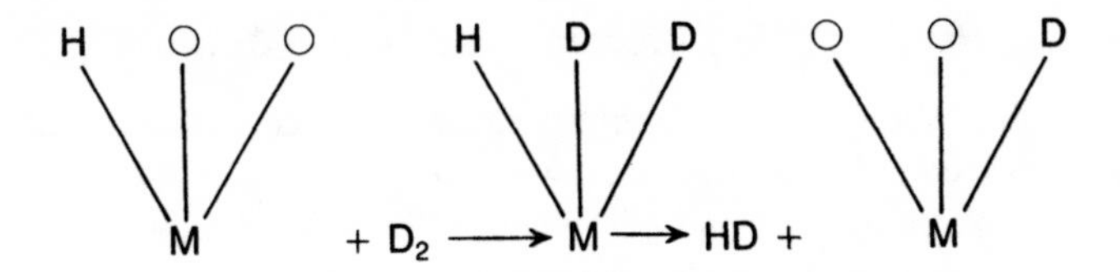

If the alternate mechanism shown below were correct, only two coordination sites would be needed.

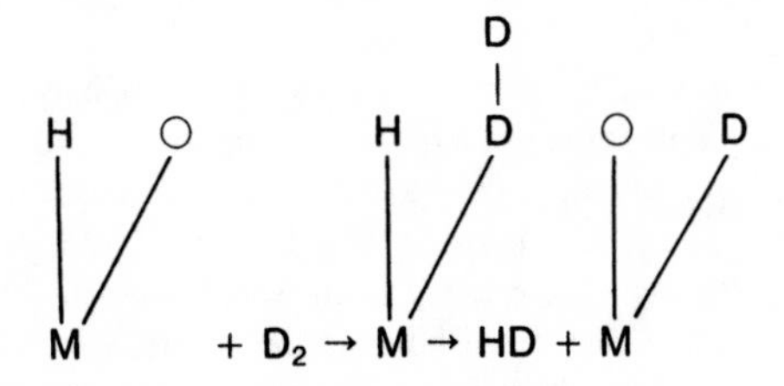

Ensemble Requirement. It seems clear that a catalyst is improved when there is a group of contiguous surface atoms present. By diluting a catalyst with an alloy, one can get an idea of how many atoms are needed. For example, to split ethane to methane may take 12 adjoining atoms.

Template Requirement. The template requirement describes the stereochemical requirements of the catalyst. All that is needed here is a chiral catalyst. This topic has been reviewed by Wynberg (1982).

11.1.3 Various Aspects of Heterogeneous Catalysis

We now discuss briefly various ways in which catalyst preparation can be scientifically controlled.

A. Drying the Catalyst Mixture for Optimum Surface

The catalyst is usually a mixture of substances. We need to have the catalyst surface at a maximum, and formed under controlled conditions, to have the catalyst components present on the surface in a reproducible state. Herrington (1982) shows how to dry the slurry to obtain a particular shape with a hollow inside and consequently a large surface which does not readily sinter. This shape, called an amphora, has been used by Sohio Research in many processes. The drying process is shown in Figure 11.1. Note the effect of the current of air in this process.

B. Bringing Reactants in Different Phases into Contact

Substances called *phase transfer catalysts* cause the mixing of two reactants that formerly did not combine because they were in different phases. A brief review of this topic is given by Reuben and Sjoberg (1981). The following example is used by these authors to illustrate how these substances work.

> Suppose we wish to esterify *p*-nitrophenol with acetic anhydride.
>
> $$O_2NC_6H_4OH + Ac_2O \rightarrow O_2NC_6H_4OAc + AcO^-Na^+$$
>
> If we mix the ingredients, no reaction occurs. Acetic anhydride is electrophilic and needs a negative charge on the hydroxyl group to attract it. Therefore we try again with the sodium salt of *p*-nitrophenol. After dissolving this substance in aqueous alkali, we add acetic anhydride, which is then hydrolyzed to acetic acid. In order to keep acetic anhydride away from water, we dissolve it in methylene chloride and try again. Once more no reaction occurs (except in concentrated solution), this time because there is insufficient contact between the phenoxide ions in the aqueous layer and the acetic anhydride in the organic layer. The yellow phenoxide layer remains in the aqueous phase. If we add catalytic amounts of tetrabutylammonium hydrogen sulfate, however, the quaternary ammonium ions pull the phenoxide ions into the organic phase as ion pairs and so reaction takes

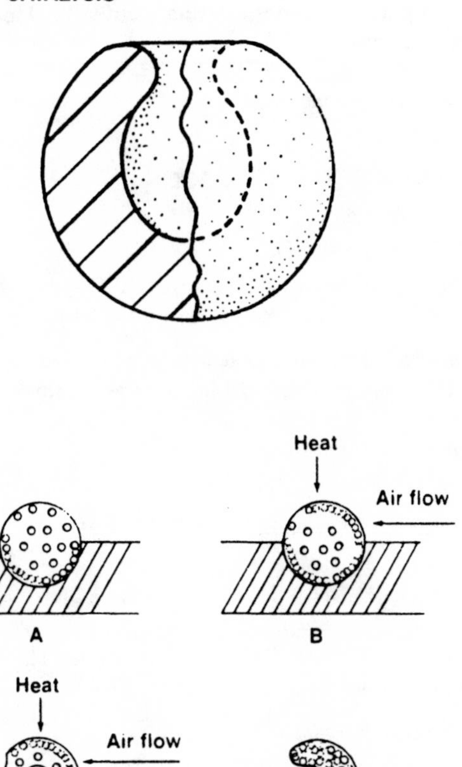

FIGURE 11.1 Method for increased area on catalyst surfaces: Preparation of amphora particles. (From *CHEMTECH* **12,** 42 (1982). Reproduced by permission of D.R. Herrington, Sohio Chemical Co., Cleveland, OH.)

place. The quaternary ammonium ions subsequently return to the aqueous phase (now paired with acetate ions) to repeat the process with other phenoxide ions until all the *p*-nitrophenol is esterified. The quaternary ammonium compound is called a phase transfer catalyst.

11.2 HOMOGENEOUS CATALYSIS

In the preceding set of catalysts the reaction occurs on the surface of the material. We now discuss the so-called homogeneous catalysts, which are in solution.

TABLE 11.2 Examples of Homogeneous Catalysts Used in Industrial Processes

1. *Acetic Acid Synthesis* (two methods)

a. $CH_3CHO \rightarrow CH_3COOH$ or CH_3COOOH

Catalyst: Mn or Co acetate.

b. $CH_3OH + CO \rightarrow CH_3COOH$

Catalyst: $RhCl_3 \cdot 3H_2O$ with an iodine co-catalyst. Cobalt carbonyl is also often used but a higher temperature is needed. See Section 4.2.4.B.

2. *Olefin Isomerization* (gives terminal double bond needed for adiponitrile formation, and for linear aldehydes to be favored in hydroformylation).

Butadiene → adiponitrile

$$CH_2{=}CHCH{=}CH_2 + HCN \overset{(A)}{\rightleftharpoons} CH_3CH{=}CHCH_2CN \overset{(B)}{\rightleftharpoons}$$

$$CH_2{=}CHCH_2CH_2CN + HCN \rightarrow NC(CH_2)_4CN$$

Catalyst $Ni[P(OAryl)_3]_4$ causes HCN reaction (A) and also isomerization (B). See Parshall (1980, p. 70) and Section 4.4.3.D.

3. *Aldehydes by Hydroformylation*

Hydroformylation (mentioned in Section 4.3.4)

$$CH_3CH{=}CH_2 + H_2 + CO \rightarrow CH_3CH_2CH_2CHO$$

Catalyst $HCo(CO)_4$ favors isomerization (C) as well as the hydroformylations(D). See Section 4.3.4.

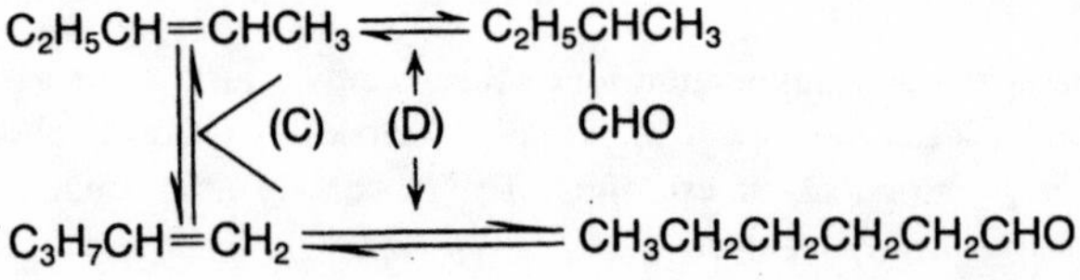

$$C_2H_5CH{=}CHCH_3 \rightleftharpoons C_2H_5CH(CHO)CH_3$$

(C) (D)

$$C_3H_7CH{=}CH_2 \rightleftharpoons CH_3CH_2CH_2CH_2CH_2CHO$$

Catalyst preparation: A cobalt(II) salt or cobalt metal is treated with synthesis gas. The $HCo(CO)_4$ is the active catalyst and is the most commonly used material.

$$CO^{+2} + H_2 + CO \rightarrow Co^0 \xrightarrow{CO} CO_2(CO)_8 + H_2 \rightarrow HCo(CO)_4$$

Rhodium salts are also very active in this synthesis, functioning at a lower temperature than the cobalt materials. Prins (1980, 419–22) discusses the mechanism.

TABLE 11.2 (*Continued*)

4. *Isocyanates*

$$ArNO_2 + 3CO \rightarrow ArNCO + 2CO_2$$

This reaction is usually carried out by reducing the nitro compound to the amine, which is reacted with $COCl_2$. The correct choice of catalyst, a Pd salt or metallic Pd or Rh, brings about the reduction and carbonylation in one step. It has not quite reached the commercial stage. See Figure 4.20.

5. *Wacker Aldehyde Synthesis*

$$CH_2{=}CH_2 + O_2 \rightarrow CH_3CHO$$

This reaction has been thoroughly discussed in Section 4.2.3.A. The catalyst mixture is $CuCl_2$ plus a trace of Pd.

6. *Olefin Epoxidation by Hydroperoxides*

There are two reactions, each using homogeneous catalysis in this situation: (1) preparation of the hydroperoxide; and (2) use of the hydroperoxide to convert an olefin into an epoxide. The catalysts used for (1) are cobalt naphthenates; those used for (2) Mo compounds, especially $Mo(CO)_6$. This reaction has been discussed in Section 4.3.2.A.

7. *Olefins Made by Metathesis*

Olefins are cleaved and recombined in a specific way to give other more valuable olefinic products. This reaction has been discussed in Sections 4.1.3.J and 4.6. The catalysts used for homogeneous systems contain a transition metal compound with a non-transition metal compound, e.g., WCl_6-$C_2H_5AlCl_2$. This reaction is also carried out in heterogeneous systems, using, for example, WO_3 on high surface alumina.

8. *Terephthalic Acid from Xylene*

One reaction of several possibilities uses Cu and Mn naphthenates for the two-step reaction with *p*-toluic acid as an intermediate. It is mentioned in Section 4.7.

Another method uses *p*-xylene with O_2 and a Co salt with methanol to give dimethyl terephthalate in one step.

You will remember these from earlier chapters. Their advantage is that they have a greater specificity because the ligands can be varied, the coordination number of the central atom changed, or the oxidation state of the central atom changed. The disadvantage of these catalysts is that they are difficult to separate from the reaction mixture and so loss of expensive metal occurs. Table 11.2 gives examples of homogeneous catalysts discussed in this book.

11.3 SUMMARY

This chapter has very briefly covered catalysis, particularly for industrial processes, giving examples of chemical types of heterogeneous and homogeneous catalysts needed for particular types of reactions.

REFERENCES

Hegedus, L. L., and J. J. Gumbleton, *CHEMTECH*, **10,** 630 (1980).

Herrington, D. R., *CHEMTECH*, **12,** 42 (1982).

Parshall, G. W., *Homogeneous Catalysis*, Wiley-Interscience, New York, 1980.

Prins, R., "Reaction Mechanisms in Homogeneous Catalysis," in R. Prins and G. Schuit, Eds., *Chemistry and Chemical Engineering of Catalytic Processes*, Martinus Nijhoff/Dr. W. Junk, Dordrecht, The Netherlands, 1980.

Reuben, B., and K. Sjoberg, *CHEMTECH*, **11,** 315 (1981).

Sachtler, W., *CHEMTECH*, **13,** 434 (1983).

Satterfield, C. N., *CHEMTECH*, **11,** 618 (1981).

Taylor, K. C., *Catalysis: Science and Technology*, **5,** 119 (1984).

Wynberg, H. *CHEMTECH*, **12,** 116 (1982).

EXERCISE

List the catalyst and catalyst type (e.g., heterogeneous dehydrogenation catalyst) used in the following syntheses:

$$\text{Benzene} \rightarrow \text{nitrobenzene}$$

$$\text{Nitrobenzene} \rightarrow \text{aniline}$$

$$CH_4 + H_2O \rightarrow CO + 3H_2$$

$$CH_2{=}CH_2 + H_2 + CO \rightarrow CH_3CH_2CHO$$

$$CH_2{=}CH_2 + 2HCl + 0.5O_2 \rightarrow ClCH_2CH_2Cl$$

$$CH_2{=}CH_2 + 0.5O_2 \rightarrow \underbrace{CH_2CH_2}_{O}$$

$$CH_3OH + 0.5O_2 \rightarrow HCHO + H_2O$$

$$C_6H_5CH_2CH_3 \rightarrow CH_6H_5CH{=}CH_2$$

$$C_6H_5CH_2CH_3 + O_2 \rightarrow C_6H_5CH(OOH)CH_3$$

$$\text{Cyclohexane} + O_2 \rightarrow \text{cyclohexanone} + \text{cyclohexanol}$$

12

Careers for Which an Industrial Chemistry Background Is Needed

This book provides a basic background for persons aiming for a future career in chemical industry or related fields. In this chapter we outline some of the types of positions that require such knowledge. This material is also aimed at some who might not realize that they need it. As the chapter is examined, this latter point will become clear.

The author has spent most of his career with a medium-sized organic specialty chemical manufacturing plant, carrying out process development work. The series of jobs he will describe are those he has observed in this company that *require* the sort of background given in this book.

12.1 CAREERS IN A MEDIUM-SIZED CHEMICAL COMPANY

The company we discuss here is involved in the preparation of products for sale to other companies, or, to some extent, to consumers. Its operative functions are as follows:

Research
Process development
Production
Sales
Technical service
Safety
Chemical analysis for each division
Application testing for each division
Technical management
Financial management

To some extent these positions have already been described in Chapter 5, but here the point of view is that expressed by the applicant's question: Just what will I be doing if I accept a _____ position in _____ company?

12.1.1 Research Chemist

Research in industry is really rather similar to that in the academic laboratory, except that its direction must eventually lead to a product or a function to be sold in order for the company to have a source of income. To bring this about, it is essential to understand the necessity of having groups of people to complete a project. One person cannot discover a cure for cancer and put it on the market. The chemical industrial groups have to understand that there are constraints on their work because of the problems faced by other groups in the same company working on other aspects of the same projects. This should be realized when they are planning their work.

As an example of this, the author's company is involved in products used as additives for the stabilization of various plastics against degradation by ultraviolet light. It is clear that the research personnel must formulate sound research programs to come up with a substance to prevent or reduce this damage. Before this, however, the chemistry of the degradation has to be understood. These basic steps are only a part of the chemist's job. A review of the literature and of what products are already being sold for this purpose is necessary to see what improvements are needed.

The research chemist will submit to the project leader (or technical or research director in a small company) proposals for carrying this out. The two people will decide whether the work is justified based on the time and the number of personnel needed. The applicant may wonder whether the reviewer will regard his work unsympathetically, which might be so if the other person had insufficient background. Today, however, companies usually make sure that research and technical directors do have adequate training and background, for the cost of overlooking precious bright ideas is too great.

The judgments so far may seem reasonable to the applicant, but the research plans are not ready yet for the stamp of approval. The financial and managerial staff will get involved in approving a major research effort. This is necessary because corporate judgment is involved in total projected sales, necessary raw materials, patent protection, available manufacturing equipment, and so on. There will also be an evaluation of this projected work as compared with other projects for which the research personnel might be needed. Usually management involvement is not too great at this early stage.

So what really happens is that the research goes along with as much depth as in the academic environment. It should be emphasized that there is an obligation to keep the management informed of the status of the project by some regular reporting schedule.

It is essential that the research chemist understands the problems that can befall projects of this sort at a later stage and be sensitive to these pitfalls. His opinion may be sought if, for instance, an analytical or application test on material produced later on presents a problem dealing with some aspect of the basic chemistry involved.

12.1.2 Analytical Research Chemist

Since the research program is dealing with a new compound series, the analyst has challenging responsibilities because it is essential that a comprehensive method be developed for the analysis for the product. It is at this stage that the product is being tested—in all sorts of mixtures if it is an additive or in body fluids if it is a pharmaceutical. Analytical research is an excellent career for many reasons. The chemist will find that all sorts of equipment is purchased, because time is money, and industries cannot wait any longer than necessary for results. Every aspect of a product's chemistry is of interest, and the analyst is supplying the critical data and is involved with discussions about their meaning.

12.1.3 Process Development Chemist

After a chemical substance has been found to have the qualities of a potential product, the process development people take over. This career has already been described in Chapter 5. The need here is to be familiar enough with the process to be able to run it safely in large scale and to get a firm picture of the economics. The process development chemist has to understand the needs of a production process, interpret the analytical results, and modify or even change the process to meet the requirements.

12.1.4 Analytical Development Chemist

In addition to being able to follow the product analysis, it is important to be able to analyze for all significant by-products in the reaction mixture. It is important to have the analytical picture on how the application tests for the product are affected by all expected by-products. This will be carried out while the process is being finalized. The impurity level will be determined through chemical and economic will be considerations. The immediate task in starting the project is to work out an analysis of the reaction mixture for appearance of product and by-products and disappearance of raw materials. The test should use simple techniques, because it should be possible to run it alongside the reactor with simple equipment.

12.1.5 Production Chemist

Not many college students aim for a career in production, probably because this type of job is seen as routine. However, there is a type of challenge that a production chemist has to meet. The laboratory chemist will have seen that all chemical reactions are subject to a large number of variables: temperature, pressure, molar excess of reagent, and so on. It is difficult to control conditions so that the duplicate goes exactly the same as the test run. In addition to the typical laboratory difficulties encountered, control of large commercial batches also involves variations in raw material from different suppliers, having two or three different people work on the same batch at different times of the day, failure of the equipment for some reason, and many other factors mentioned in this book. A well-qualified chemist is needed to be alert to the possibilities when unexpected events occur and to take the appropriate action. There are also problems of what to do when a batch of material is substandard. To deal with such problems, the production chemist has to understand fully the chemistry of the processes he is dealing with. He is expected to recommend process and equipment changes based on his experience with the process.

The production chemist should enjoy dealing with people and gain satisfaction from turning out large quantities of quality material with the minimum of disruption and at low cost.

12.1.6 Foreman

The foreman meets many of the challenges of the production chemist. Even though he may have only one or two processes under his control, he needs a good chemical background. Foremen are often promoted from operator status, where the control is carried out through a meticulously written process. However, some chemistry courses should be taken, for example, at the community college level. One reason he needs chemistry is that he should be able to explain to the operators what they are doing.

12.1.7 Control Lab

The sort of person who would be happiest in the control lab is one who would be able to repeat many analytical procedures over and over again with complete reliability. This is a perfect job for an ambitious new college graduate because this post can be used as a first step upward by seeing what tests are carried out on the various products and becoming familiar with the latest analytical equipment. This is also an excellent job for a high school graduate who does not plan any further schooling.

12.1.8 Director of Safety, Health, and Ecology

It is a mistake to leave responsibility for safety, health, and ecology functions to people with production responsibility. This is because these factors may be ignored under the pressure of meeting a production schedule. There are three related aspects to this type of work, each of which involves a considerable background in chemistry. One deals with protecting the personnel on the job, keeping the exposure to toxic chemicals below the value that causes harm. Another concerns handling chemicals so that no dangerous reactions, decompositions, explosions, or fires occur. There may also be ecological problems in the neighborhood because of the plant, which involve similar considerations. Today these considerations involve management level decisions. The person fitting this job must be able to understand the chemical factors causing these problems, and assemble chemically trained personnel to take action to solve them.

A. Safety Engineer

There must be a person at a plant who is charged with carrying out company safety policy. As part of his job he assembles a safety data sheet with data supplied by many people in the organization. This data sheet collects all the information on every chemical used in the process in a form readily available in an emergency. Such a document is described in Section 5.1.7.

12.1.9 Sales

Sales is a career very few students plan for if they major in chemistry. It is a fact, however, that it represents a job that is a challenge and a rewarding career for a well-trained chemist. Most chemical substances manufactured are for sale to other companies, through an activity called marketing. In negotiating a sale between two such institutions, price is only one of many concerns. An example of a topic that might be discussed is the question of whether a higher specification product at the same price is possible. Although the final answer is, of course, the responsibility of management and the production department, the sales person should have a good idea of whether this might be possible and be able to converse intelligently with the other representatives. The best background for this career is for the new employee to be introduced to the company as a development chemist and to see by experience the many factors that affect product cost.

12.1.10 Technical Service

It is as necessary for a technical service representative to have a good background in chemistry as it is for a sales person. Technical service tries to solve the purchaser's problems by visiting the plant where the company's product is being used. He has to understand how his company's product is prepared and whether it is causing the problem. Also, he has to carry back to his people the needs of the customer: a purer product, a modified structure. He has to understand the chemistry of the product as it is used by the customer.

12.1.11 Management Positions

When a scientist advances in a company, he often has to decide whether to stay in laboratory-related work or take on administrative work. It is sometimes said that the latter choice means giving up the work that he has been trained for. This is certainly not true today. All the management people who run chemical companies use their chemical knowledge every day. It would be foolhardy indeed to have a person in charge who could not talk to the technical staff on terms they can each understand.

12.1.12 *Opportunities in Chemistry*—A Report Written about Government, Industrial, and Academic Careers

In turning to review chemical careers in a general sense, it is essential to refer to a remarkable report called *Opportunities in Chemistry*, compiled by a large committee of eminent chemists chaired by Pimentel (1985). This document is highly recommended for those approaching graduation at the undergraduate level and for graduate students. It covers in a form readable by those with this background the following topics:

1. Current advanced research areas in chemistry.
2. The ways that chemistry can be utilized for social needs.
3. The resources to be investigated to carry out the above.

The report is therefore suitable for those considering an industrial, government, or academic career.

The author also recommends the report *Chemistry in the Economy*, by Harris and Tishler (1973) (see Chapter 1). Here eminent scientists from industry describe their companies and fields of endeavor.

12.2 GOVERNMENT POSITIONS

Today federal, state, and municipal governments have many positions requiring a knowledge of chemical products and how they are produced. Many departments of the federal government have a deep involvement with chemical related problems, for example, defense, health and human services, and agriculture. The various National Institutes of Health represent sources of chemical careers at all scientific levels. Regulatory agencies throughout the country also employ people to run tests and make scientific decisions affecting our lives. State and even many municipal governments should be filling posts with chemically trained people for regulatory staff positions of all sorts. Also, those holding elective posts are often confronted with decisions such as whether to allow a company to locate in a certain area. How much better it would be for these people to have a general ideal of what is involved in this situation.

12.3 CAREERS IN THE PHARMACEUTICAL INDUSTRY

Pharmaceutical positions will, of course, include those careers already mentioned in the description of the medium-sized chemical company. In addition, there should be listed the following:

12.3.1 Careers in Biological and Medical Testing

Workers in biological and medical testing should be alert to the possibilities of the effect of process changes on the product. It will make their job easier when they are testing products made on a large scale, that they are alert to what changes in biological and medical effects can occur when the impurity spectrum and physical form of the material changes.

12.3.2 Careers in Formulation

Products in the pharmaceutical industry are never sold as the pure active ingredient. The manufacture of pills, sometimes called drug delivery systems, requires skill. The pill ingredients have to be chosen to release the active ingredient in the diseased organ of the body. The active ingredient has to be in the correct crystal form. Organic chemicals often crystallize in a variety of polymorphic forms, even though all are equally chemically pure. The formulator has to be aware that process changes in manufacture can cause another crystal form to be isolated that can result in the pills sintering. He must be aware that if the research team requires a new dosage form such as a salt or a derivative, he must find a new drug delivery system. He has to have a procedure to have an extremely small amount of active ingredient in each pill and must be aware of the difficulty of manufacturing extremely uniform lots of chemicals.

12.4 CAREERS IN THE SOAPS AND DETERGENTS INDUSTRY

The chemistry of soaps and detergents was touched on in the large-scale processes discussion in Chapter 10. The list of chemicals in this industry group has become quite lengthy since the time when we washed our clothes in the sodium salts of fatty acids. The surfaces to be cleaned now include the synthetic fibers and plastics as well as coated cotton fabrics. Soaps are now expected to work well in hard waters. Surface active agents and water softening agents are mixed into the soaps. Interesting careers for all academic levels can be found here.

12.5 CAREERS IN THE PERSONAL CARE PRODUCTS INDUSTRY

Interesting chemical careers are to be found in the preparation of products that people apply to their hair, skin, and nails. The products include hair sprays, shampoos, hair coloring, nail lacquers, suntanning preparations, deodorants and antiperspirants, and dentifrices. There is a wealth of chemistry needed in the manufacture of these substances.

12.6 CAREERS IN THE FERTILIZER INDUSTRY

We have already discussed three of the chemicals used in this industry: ammonia, sulfuric acid, and phosphoric acid. Formulated nitrogen, phosphorus, and potassium preparations are needed to replenish the soil. These major nutrients are supplemented with the secondary nutrients calcium, magnesium, and sulfur as well as the micronutrients boron, copper, iron manganese, molybdenum, zinc, and possibly chlorine, silicon, and sodium. The chemistry of the manufacture of various preparations of these materials in the form where the nutritional metal is in the form most available to the plant (usually an organic metal chelate) presents a challenge to the fertilizer industry. The chemists involved are concerned with the economic aspects of the processes for two reasons. First, the cost of the agrochemicals used today is rather high even considering the amazing productivity increase that has resulted. Second, the price of these materials as far as third-world countries are concerned practically prohibits their use. For the latter situation particularly, cheaper highly active fertilizer forms are needed.

12.7 CAREERS IN THE PESTICIDE INDUSTRY

Pesticides are used in agriculture in three ways: for the control of insects, weeds, and plant diseases. The processes for a few of these products have been mentioned in Chapter 10. Products are also used in nematode and rodent control and as plant growth regulators. The chemistry and biochemistry of this enormous industry represents a tremendous opportunity for graduate chemists.

12.8 CAREERS IN THE FOOD PROCESSING INDUSTRY

There are two reasons for the use of chemical products in the food processing industry. The first is the preservation of food against spoilage by bacteria and degradation of the many valuable components such as proteins, fats, carbohydrates, vitamins, and minerals. Calcium propionate is one of a huge number of products in the field. The second reason is the formulation of foods with baking powder and other materials as mixes to make the process of cooking much easier and shorter.

Examples of methods of preservation of foods are by dehydration and freezing. Biochemists have made the necessary studies of enzyme-caused changes in foods and have found out how to inactivate them by heat treatment or by additives. Another example of processing food is by injecting papain into an animal just before slaughter as a tenderizing treatment. Another example of food treatment is in the processing of grain for making ready-to-eat cereals. Ground whole grain where the outer husk has been removed deteriorates rapidly. Treatment of the grain to prepare a nutritious cereal takes much chemical and biochemical know-how.

The preparation of edible oils and margarine and the development of food additives for stability and vitamins for nutrition are other ways in which chemists and biochemists are deeply involved.

12.9 OTHER BRANCHES OF THE CHEMICAL INDUSTRY

The aim of this chapter has been to awaken you to the many careers waiting for chemists. The following is a list of some branches of this large industry that have not yet been mentioned:

1. Plastics and resins
2. Textile fibers
3. Natural and synthetic rubber
4. Protective coatings

These are based on polymerization processes. The products are used for the electronic, electrical, and paint industries.

Four other branches of industrial chemistry could be covered, especially if a manufacturing facility is in the area. These are

1. Nuclear energy
2. Pulp and paper
3. Ferrous metals
4. Glass

12.10 NEEDS OF THE GENERAL PUBLIC

Much has been written about the value of a university education today. A small but significant part should be devoted to courses that show how science has an impact on society. The general public needs training on what will be the result of voting to accept a manufacturing plant in a community. As the result of taking this course, people will have a more informed attitude toward a chemical manufacturing plant in their community and will be able to talk to officials in a more reasonable way. Newspeople have often been guilty of causing a hysterical response to a perceived threat from a chemical plant and could benefit from becoming more informed about the chemical industry. The Pimentel report (1985) discusses this question.

12.11 WRITING CHEMICAL PROCESSES AND REPORTS

An important but often overlooked part of all the careers discussed earlier concerns reporting the assembled data. Usually the scientist must write a report at a stage where, if more time were available, additional experiments could be performed to make the data more secure. So he or she has to evaluate the data on hand and write the report so that the reader can understand it and act on it even though he or she will likely have a different background from the writer.

To be a success in any of the careers described above, the scientist has to be able to do this necessary task.

We now present an outline of the basic parts of a chemical process to show how to organize the data from many people in the organization into a comprehensive yet succinct document.

When you want to find out about what goes on in a chemical manufacturing establishment, you should consult the processes that have been written and are being used. If they are up-to-date, and represent what is being currently carried out, and are properly formulated, the documents will supply all the information needed as a starting point for a wide variety of organizational needs.

We now list the basic parts of a chemical process which should be in written form in one place when the process is being carried out on a commercial scale. If this is done, it will ensure that the production will be carried out in a satisfactory manner, and that the other people involved, such as development chemists, analysts, purchasing agents, financial people, environmental and engineering personnel, industrial hygienists, and management can find out basic facts.

The process should include:

1. Historical summary of the chemistry leading to the present process and a description of the product's end use.
2. Summary of process.
3. Ingredients needed for the batch or for 1000 kg of product. Raw material specifications are essential.
4. Equation.
5. Equipment flow diagram for every part of process, including all environmental monitoring and protection set-ups.
6. Stepwise procedure (by each step, caution statements are inserted on safety in handling equipment and chemicals).
7. Analysis procedures needed at critical points for items 3, 6, 8, 9, 10, 11, and 12.
8. Yield and specifications for the product.
9. Recovery procedures for solvents and second crops of crystals; yield expected.
10. Weights of solid, liquid, and gaseous wastes. Detailed procedures for handling all of this material.
11. Safety data sheets listing all handling problems concerning toxicity, fire, chemical instability, etc., of reagents, products, and by-products.
12. First aid procedures in case of accidental exposure to all reagents, products, and by-products.
13. Storage information for raw materials and products.

Supervisors should have a copy of the cost calculation for their basic background information.

This list is included in this introductory book for the prospective employee so that he or she can see the many interrelated steps involved in order for the company to prepare the materials in an economical, safe manner, and how necessary it is for each person contributing to see what the process entails.

Some of this information will be in summary form. For example, there would be analyses described by title stating why the analysis is carried out and what the results show.

Clearly a succinct style is needed, because the material will be read and used by a variety of people with different kinds of training.

12.12 FINAL REMARKS ON INDUSTRIAL CHEMISTRY

The reader of this book should now be able to understand the essential characteristics of an industrial chemical process, and how it differs from a laboratory procedure. To review what has been learned, one can check examples of reactions using unconventional catalytic procedures which have been discussed in this book, and are listed in the index under "Catalytic processes, air oxidation". Here, air oxygen is incorporated into the product or causes reagent recovery.

The pressure to find these processes has come from economic and environmental concerns. Some success has come with the petrochemicals, prepared in enormous quantities. This approach needs to be used on a much wider scale, and offers many career opportunities for people in this field.

12.13 SUMMARY OF THE BOOK

This book is a brief summary of the chemical industry, emphasizing an overview of inorganic, fermentation, and organic processes. You should now know what to expect when you enter any chemical industry career, having been introduced to the sort of things that the people talk about on a daily basis, such as equipment, costs, and scale-up. Also, you will not be leaving your detailed theoretical chemistry behind, since the most successful chemists are those who can interpret the needs of an industrial process most clearly, that is, understand the chemistry involved and take chemically based actions to solve the problems. The book will serve as a basis for an introductory polymer course.

REFERENCES

Harris, M., and M. Tishler, *Chemistry in the Economy*, American Chemical Society, Washington, D.C., 1973.

Pimentel, G. C., *Opportunities in Chemistry* (Committee Report), National Academy Press, Washington, D.C., 1985.

References for Supplementary Reading

This book has been written for a person who wants an overview of the chemical industry, to obtain a general impression of what he or she will need if a chemical career is chosen.

It is recommended that the following list of books be on hand if the course is to be presented by a university or industrial concern. These have been referred to at several places in the chapter references.

Kent, J. A., *Riegel's Handbook of Industrial Chemistry*, 8th ed., Van Nostrand-Reinhold, New York, 1983.

Thompson, R., *The Modern Inorganic Chemicals Industry*, The Chemical Society, London, 1977.

Weissermel, K., and H.-J. Arpe, *Industrial Organic Chemistry*, Verlag Chemie, Weinheim, 1978.

Wittcoff, H., and B. G. Reuben, *Industrial Organic Chemicals in Perspective*, Part 1, Wiley-Interscience, New York, 1980.

In addition, the following references are given for those who wish to explore further the topics mentioned in the book. This short list will enable the student to take a first step toward getting more background in a particular topic.

General Reference for All Topics

Kirk, R., and D. Othmer, *Encyclopedia of Chemical Technology*, 3rd ed., Wiley, New York (18 volume set).

All of the chemical industry is covered by authorities in the individual areas.

Inorganic Chemicals

Hocking, M. B., *Modern Chemical Technology and Emission Control*, Springer-Verlag, Berlin, 1985.

This is a current review of some of the Thompson topics covered in greater depth, and also petroleum refining. Emission control equipment is described.

Thompson, R. (see above).

In addition to the topics covered in this book, chemicals from fluorine, bromine, boron, and titanium, as well as silicates, peroxides, and products from the nuclear fuel industry are reviewed.

Fermentation Chemicals

Kent, J. A., (see above).

One chapter is devoted to this topic.

Peppler, H. J., and D. Perlman, *Microbial Technology*, 2nd ed., Academic, New York, 1979.

This is a two-volume work in which various authorities discuss topics such as: Microorganisms for Waste Control; Economics of Fermentation Processes; Microbial Production of Amino Acids, Organic Acids, and Antibiotics; Vinegar, Wine, and Beer Production.

Organic Chemicals

Shreve, R. N., and J. A. Brink, *Chemical Process Industries*, 4th ed., McGraw-Hill, New York, 1977.

An authoritative review of many inorganic and organic processes emphasizing detailed process diagrams.

Weissermel, H., and H.-J. Arpe (see above).

A comprehensive review of the organic chemical industry. The discussion of the chemistry of the processes is quite detailed.

Wittcoff H., and B. G. Reuben (see above).

A succinct yet complete review of the whole organic chemical industry. The coverage is more comprehensive than this text, and emphasizes economic factors controlling the choice of processes.

Scale-up

Bisio, A., *Scaleup of Chemical Processes: Conversion from Laboratory Scale Tests to Successful Commercial Size*, Wiley, New York, 1985.

Clausen, C. A., and G. Mattson, *Principles of Industrial Chemistry*, Wiley-Interscience, New York, 1978.

Cost Calculations

Clausen, C. A., and G. Mattson (see above).

Holland, F. A., F. A. Watson, and J. K. Wilkinson, *Introduction to Process Economics*, 2nd ed., Wiley, New York, 1983.

Environmental Impact of a Process

Ahmad, Y. J., *Economics of Survival: The Role of Cost–Benefit Analysis in Environmental Decisions*, United Nations Environmental Programme, Nairobi, 1981.

Cottrell, A. H., *Environmental Economics: An Introduction for Students of the Resource and Environmental Sciences*, Halsted Press, Wiley, New York, 1978.

Neely, W. B., Chemicals in the Environment: Distribution Transport. Fate. Analysis, Dekker, New York, 1980.

Stenhouse, J. I. T., "Pollution control" in *Chemical Engineering and the Environment*, A. S., Teja, Ed., Halsted Press, Wiley, New York, 1981.

Air and water pollution control methods and equipment are discussed in Hocking (see above).

Equipment

Chen, N. H., *Process Reactor Design*, Allyn and Bacon, Boston, MA, 1983.

Distillation

Nisenfield, A. E., and R. C. Seeman, *Distillation Columns*, Monograph Series, Instrument Society of America, Research Triangle Park, North Carolina, 1981.

This book gives the basic background needed for understanding this equipment.

Extraction

Lo, T. C., M. Baird, and C. Hanson, *Handbook of Solvent Extraction*, Wiley Interscience, New York, 1983.

Crystallization

Van Hook, A. J., *Crystallization: Theory and Practice*, Reinhold, New York, 1961.

Walton, A. J., *The Formation and Properties of Precipitates*, Wiley-Interscience, New York, 1967.

Pesticides

Buchel, K. H., *Chemistry of the Pesticides*, Wiley, New York, 1983.

Pharmaceuticals

Burger, A., *A Guide to the Chemical Basis of Drug Design*, Wiley-Interscience, New York, 1983.

Lednicer, D., and L. A. Mitscher, *The Organic Chemistry of Drug Synthesis*, Wiley-Interscience, New York, 1977.

Wolff, M. E., *Burger's Medicinal Chemistry*, 4th ed., in three volumes, Wiley, New York, 1979–1981.

Catalysts

Leach, B. E., *Applied Industrial Catalysis*, Academic, New York, 1983.

A work in three volumes, with chapters by specialists on many topics, for example, Fischer–Tropsch synthesis, and methanol synthesis.

Index

Printed in the United States
98917LV00002B/81/A

9 780471 826576